Modulation in Electronics and Telecommunications

Edited by George Dekoulis

Published in London, United Kingdom

IntechOpen

Supporting open minds since 2005

Modulation in Electronics and Telecommunications
http://dx.doi.org/10.5772/intechopen.75319
Edited by George Dekoulis

Scientific Contributor Anatoliy Platonov

Contributors
Sergey Rassomakhin, Ashish Singh, Akhilesh Kumar Pandey, Yufeng Tao, Ramón José Pérez Menéndez, Ichihiko Toyoda, Anatoliy Platonov, Ali Reza Hazeri, Krishnananda Shet, Durga Prasad, Mohammad Aneesh

First published in London, United Kingdom, 2020 by IntechOpen
IntechOpen is the global imprint of INTECHOPEN LIMITED, registered in England and Wales, registration number: 11086078, 5 Princes Gate Court, London, SW7 2QJ, United Kingdom
Printed in Croatia

British Library Cataloguing-in-Publication Data
A catalogue record for this book is available from the British Library

Additional hard and PDF copies can be obtained from orders@intechopen.com

Modulation in Electronics and Telecommunications
Edited by George Dekoulis
p. cm.
Print ISBN 978-1-78985-489-3
Online ISBN 978-1-78985-490-9
eBook (PDF) ISBN 978-1-83968-965-9

For EU product safety concerns:
IN TECH d.o.o., Prolaz Marije Krucifikse Kozulić 3, 51000 Rijeka, Croatia, info@intechopen.com or visit our website at intechopen.com.

Meet the editor

Prof. George Dekoulis received his PhD in Space Engineering and Communications from Lancaster University, UK, in 2007. He received a First Class BEng (Hons) degree in Communications Engineering from De Montfort University, UK, in 2001. He has received several awards from the Science and Technology Facilities Council (STFC), the Engineering and Physical Sciences Research Council (EPSRC), and the Hudswell International Research Scholarship (IET). He is currently a professor at the Aerospace Engineering Institute (AEI), Cyprus. He is a founder of the IEEE Aerospace and Electronic Systems Society (AESS), Cyprus, and was the general chair of IEEE Aerospace Engineering Innovations 2019 (IEEE AEI 2019), Limassol, Cyprus. He has previously worked as a professor in Aerospace Engineering at various departments, teaching such subjects as space and planetary physics, aeronautical and space engineering, professional flight, robotics/mechatronics and mechanical engineering, computer science and engineering, and electrical and electronics engineering. His research focuses on the design of reconfigurable aerospace engineering systems.

Contents

Preface

This book presents information on modulation and examines methods for improving the deployment and technical characteristics of wireless communication systems and networks.

Chapter 1 shows that low-energy transmitters with adaptive amplitude modulation (AM) may transmit narrowband signals without coding in real time and with spectral–energy efficiency, attaining the limits established by Shannon's theorems. The only requirement is high-quality feedback channels delivering controls from base stations to the forward transmitter.

Chapter 2 presents an original and low-complexity method for fast and accurate analysis and demodulation of orthogonal frequency-division multiplexing (OFDM) signals with unknown structure and parameters in conditions of uncertainty and noise. The focus is on identifying the type of modulation and parameters of the subcarrier as well as eliminating inter-channel interference and phase errors.

Chapter 3 presents research on the capabilities and advantages of electro-optic lithium niobate (LiNbO3) modulators for optical communications. It discusses methods of fiber integration, connection to microwave electrodes, box sealing, and fabrication for on-chip implementation. In addition, it presents results of interferometric optical sensing via an electro-optic manipulating light field.

Chapter 4 considers possibilities of using polarization of radio frequency (RF) signals and propagation direction as additional carriers of information. Attention is paid to the most important aspects of the proposed modulation schemes: polarization-agile antennas and active integrated array antennas, which integrate an oscillator and modulators, and polarization discrimination antennas. The chapter also introduces receiving antennas.

Chapter 5 discusses the main optical phase-modulation techniques, and the most appropriate technique is used for the elaboration of original phase modulation processes for the optical Sagnac interferometer gyroscope. It is shown that closed-loop configuration of gyroscope provides a greater accuracy (sensitivity to declinations) and scale-factor stability than the open-loop analog.

Chapter 6 presents the results of studies on the analysis and design of injected oscillators generating carrier signals. New results concern the analytical methods of locking and pulling phenomena. The chapter also reviews the application of injected oscillators, in particular, injection-locked frequency dividers. It also proposes an original method of enhancing locking range.

The final chapter contains a review of dual-band, multiband, and ultra-wideband antennas. The review classifies and compares antennas according to antenna feeding and loading of antennas using slots, notch, and coplanar structures. It also discusses the basic geometry of patch antennas along with their equivalent circuit diagram and particularities of their applications.

Anatoliy Platonov, DSc, PhD.
Professor,
Institute of Electronic Systems Warsaw,
Warsaw University of Technology,
Poland

Chapter 1

Perfect Signal Transmission Using Adaptive Modulation and Feedback

Anatoliy Platonov

Abstract

The research results show that adaptive adjusting of modulators over feedback enables development of the "perfect" communication system (CS) transmitting analog and digital signals in real-time without coding with a bit rate equal to the forward channel capacity and limit energy spectral efficiency. These and other feasibilities unattainable for known CS are the result of transition from the direct transmission of samples of the input signal to the transmission of sequences of their estimation errors cyclically formed at the input of forward transmitter (FT) modulator. Each transmitted error is formed as a difference between the value of input sample and its current estimate computed in the receiver in previous cycle and delivered to FT over feedback. Growing accuracy of estimates decreases estimation errors and permits their transmission permanently increasing the modulation index and maximizing the amount of information delivered to the receiver. Unlike CS with coding, adaptive feedback CS (AFCS) can be optimized using Bayesian estimation and information theory. Absence of coders simplifies the construction of FT and reduces their energy consumption and cost. Moreover, adaptive properties of AFCS permit to maintain the perfect mode of transmission in every scenario of application. The chapter presents analytical backgrounds, experiments results and research genesis including the reasons for absence of AFCS in modern communications.

Keywords: wireless, adaptive modulation, feedback, Bayesian optimization, perfect transmission, limit energy-spectral efficiency, channel capacity, Shannon's limits

1. Introduction

The publication of Shannon's fundamental works [1, 2] coincided with the appearance of the first computers and urgent need for the development of fast and reliable channels for digital data transmission. Shannon's theory led to the almost immediate development of backgrounds of the coding and digital CS theory. The parallel fast development of high-resolution AD-DA converters and digital technologies made digitizing and coding the basic principle of signal transmission.

The side effect of the successes of digital CS was initial lack of interest in CS with feedback channels (FCS) and codeless signals transmission, although this possibility was noted by Shannon in [1, 2]. The first work of Elias [3] in this direction was published seven years later, in 1956. The results of the work showed that ideal feedback channel and proper setting of the modulation gains permit to transmit analog signals without coding in real time with the limit bit rate equal to the capacity of the forward channel—the result unfeasible in digital CS with coding. Moreover, the absence of coders radically simplified the construction of FCS transmitters. Initially not noticed, this work initiated a great cycle of research in the optimization of FCS (see e.g. [4–13]) carried out in 1960s in MIT, Bell Lab., Stanford University, NASA, and other research centres. The results of these investigations unambiguously confirmed that modulation and feedback enable a development of simple CS transmitting signals and short codes in real time perfectly and with minimal distortion. Moreover, analytical results of the research determined a way to design of the perfect FCS design.

However, since the mid-1970s, interest in the research in the FCS theory sharply declined, and, during subsequent decades, only a small number of academic papers were published. The main reason was the lack of practical results, as well as a pessimistic evaluation of the entire direction of research ("The subject itself seems to be a burned out case" [14], p. 324). It is worth adding that, at that time, short-range transmission was provided by wires, and there was no special need in wireless FCS. At the same time, development of digital technologies, communications and automatics generated a lot of complex theoretical tasks, and the industry required specialists. As a result, most FCS researchers took up these tasks.

The situation changed with the appearance of mobile communications and wireless networks (WN) containing a great number of wireless end nodes (EN), each communicating with the base station (BSt) over forward and feedback channels. This renewed interest in FCS was still, however, strictly academic [15] and without any practical results. Having no alternative, currently, all the channels of WN employ only the coding principle of transmission.

Apart from the traditional requirements for CS (maximal rate, quality, reliability, range of transmission, etc.), battery-supplied or battery-less low-power transmitters of EN should be minimally complex and minimally energy-consuming, and should satisfy a large number of the other, sufficiently rigorous requirements [16] such as maximal energy efficiency of transmission, optimal utilization of the channel bandwidth, reduction of inter-channel interference, security of transmission, and others. The set of these characteristics is now defined by the general term "performance," and the main task of designers is the improvement of the systems or channels' performance.

The development of the first generations of WN and corresponding FT did not cause any particular difficulties, but each subsequent generation does pose new, increasingly complex problems. One should stress that the performance of the lower, physical (PHY) layer channels EN-BSt dramatically influences the performance of the overall network regardless of the particularities of the higher layers' organization.

The design of the PHY layer channels is carried out almost independently from that of the higher layers and software of WN, and requires thorough knowledge of mathematics, signals processing, communication and information theory, and so on. Nevertheless, even among experienced designers, "the task of changing from a cable to wireless is still seen as a daunting prospect; wireless retains its reputation of being close to black magic. For most designers, it is an area where they have very little ability to change anything, other than the output power" [16]. A similar sentiment is expressed in [17].

This is not an isolated opinion. A large number of recent publications question the capability of the modern theory to provide any noticeable improvement in wireless transmission: "Shannon limit is now routinely being approached within 1 dB on AWGN channels ... So is coding theory finally dead? ... there is little more to be gained in terms of performance [18]"; "Whether research at the physical layer of networks is still relevant to the field of wireless communications? ... any improvements are expected to be marginal [19]." Similar evaluations of the state of theory can be found in [20–22] and other works.

Analysis of the sources of problems showed that the main reason is the lack of efficient theoretical basis permitting one to investigate the behavior of wireless CS in different scenarios and to choose the versions best suited to the goals of the project.

In the following sections of the chapter, we discuss these problems and their solution. To simplify the discussion, apart from CS and FCS, we use the following abbreviations: CSC—communication system with coding and AFCS—adaptive feedback communication system transmitting signals over the forward channels (FT) using analog modulators (AMs) adjusted by the controls formed in BSt and delivered to FT over feedback channels.

2. Sources of difficulties in improvement of short-range CS performance

Let us clarify the subject of discussion. The term "performance", relatively new in communications, is broader than the term "quality". Furthermore, the evaluation of CS performance has its own groups of the criteria used to compare the systems by their general utility characteristics, further called the "performance" criteria. Another, relatively narrow and stable group of "analytical" criteria is used in the research as a tool permitting to improve one or a part of the performance criteria.

2.1 Basic performance criteria

The performance criteria determine the required, desired, or real characteristics of the future or existing wireless CS permitting to use these systems for solution of definite tasks in the given conditions, and to evaluate the corresponding benefits, costs, and risks [16]. The main criteria of this group are:

- range of reliable transmission
- data rate and throughput
- latency
- frequency range and channel bandwidth
- power and energy consumption
- security of transmission
- interference and coexistence
- resistance to industrial disturbances and changes in the environment
- possibility of supplying from renewable energy sources
- design, production, and deployment costs
- size, weight, price, and other characteristics.

These criteria are used for elaboration of standards and have no analytical tools for a prior evaluation of performance. Instead, each of the listed criteria has a fixed numerical evaluation determining the corresponding requirement to CS. Each standard defines a class of CS with a unique combination of performance criterions granting these systems the ability to solve definite tasks under definite conditions better than other systems. For a new system to have better performance, it should pass a certification which confirms the existence of new qualities. Moreover, to become the standard, it should be manufactured at least by three independent firms [16].

Analytical criteria serve a different purpose, and are used to establish conditions that allow the rate, quality, reliability of transmission, and other characteristics of CS to approach their limit or given values in different conditions and under given constraints. These criterions are built on adequate mathematical models of the main components of CS or of the system as a whole, and their values depend on all the basic factors influencing the work of the system. Some of these factors can be regulated by designers who search for their combination that either maximizes the "main" criterion (e.g. the rate of transmission) or minimizes its value (e.g., transmission errors) taking into account existing limitations. The results of research determine the approach to design the best system in a given class under given conditions and limitations.

Analytical evaluations of the quality of transmission are not used in performance criteria but the research results create a rigorous basis for the design of more efficient CSs and for their emergence within new standards. Nevertheless, RF (radio frequency) design "is typically the smallest section of any wireless standard" and "the hardware definition may be less than 5% of the total specification in terms of the number of pages ([16], p. 20).

Currently, the term "improvement of performance" is widely used in communications and pushed out also not strictly defined earlier term "improvement of quality" of the systems, channels, transmission, etc. In this chapter, we use both these terms. It is worth adding that the term "perfect" should be taken literally: the performance of the AFCS discussed below does attain the limits established by information theory. Note that this discussion clarifies not only the terminology, but also the relations between different groups of criteria mutually connected through plural tradeoffs. The analytical results presented below allow for the simultaneous improvement of several performance criteria (to which we will return in the final discussion).

2.2 Basic analytical criteria

Nowadays, commonly used analytical criteria in CSC performance include the bit rate [bits/s], energy [J/bit], and spectral [bit/s/Hz] efficiencies of transmission, as well as bit error rate (BER). Sets of possible values of energy-spectral efficiencies have upper bounds, and the task of designers is to make the characteristics of the system approach these boundaries under possibly smaller BER. As the basic references, the theory employs limit values of bit rate and efficiencies of the transmission usually computed for linear memory less channels with additive white Gaussian noise (AWGN). So, limit bit rate determines the capacity of the channel (Shannon's formula) as follows:

$$C = F_0 \log_2\left(1 + \frac{W^{sign}}{N_\xi F_0}\right) = F_0 \log_2(1 + Q^2), \tag{1}$$

where W^{sign} is the power of signal at the channel output, $N_\xi/2$ is the double-side spectral power density of AWGN, $2F_0$is the channel bandwidth, and:

$$Q^2 = \frac{W^{sign}}{\sigma_\xi^2} = \frac{W^{sign}}{N_\xi F_0} = SNR^{Ch}, \quad (2)$$

is the signal-to-noise ratio (SNR) at the channel output.

The energy efficiency of transmission ("energy per bit") $E^{bit} = W^{sign} T^{bit} = W^{sign}/C$ [J/bit] is determined as the energy necessary for transmission of a single bit of information with a bit rate equal to the capacity of the channel ($T^{bit} = 1/2C$ is the time of transmission).

The spectral (or bandwidth) efficiency C/F_0 describes a number of bits transmitted per second per 1 Hz of the channel bandwidth. The limit values of energy-spectral efficiencies and SNR are connected by the relationships [23–25]:

$$\frac{E^{bit}}{N_\xi} = \frac{F_0}{C}\left(2^{\frac{C}{F_0}} - 1\right) = \frac{Q^2}{\log_2(1+Q^2)}. \quad (3)$$

This formula directly follows from (1) and (2) and is convenient for practical applications. Another frequently used but less convenient measure of energy efficiency is defined as C/W_0 [bit/s/W], where W_0 is the power of transmitter.

The example of the expression for the BER computed for two orthogonal signals transmitting particular bits over channels with AWGN has the form:

$$\text{BER} = \text{Pr}^{bit} = 1 - 2\Phi\left(\sqrt{\frac{E^{bit}}{N_\xi}}\right), \quad (4)$$

where $\Phi(x)$ is the tabulated Gaussian integral:

$$\Phi(x) = \frac{1}{\sqrt{2\pi}}\int_0^x \exp\left(\frac{z^2}{2}\right)dz. \quad (5)$$

These relationships are mutually connected by multiple tradeoffs: power-bandwidth tradeoff; tradeoffs between BER and energy efficiency, deployment efficiency-energy efficiency, and many others (see, e.g. [23, 26]). In these conditions, the development of a regular approach for the optimization of CSC is practically impossible.

The sources of difficulties are:

i. Impossibility to find, among the infinite set of possible codes, the code minimizing errors of transmission: "The existence of optimal encoding and decoding methods is proved, but there are no methods indicated for the construction or technical realization of these results [27]."

ii. Impossibility to formulate any expressions for current (not only limit) bit rate and energy-spectral efficiencies.

iii. Both the quality of transmission and the results of CSC optimization directly depend on the scenario of the system application (placement of the system, characteristics of the environment, fading, noise, path loss, etc.). Implementation of theoretical results is only possible if there is a possibility of at least partial channels identification but a system optimal in one scenario will not be optimal in another.

iv. The lack of the regular analytical approach to optimization of CSC makes impossible evaluation of the potentially achievable bounds of transmission quality and the search for the most efficient technical solution permitting their achievement.

There are many approaches for the improvement of CSC (e.g., [28–30]) but their discussion is beyond the scope of the chapter. We will note the fact stressed in the literature (e.g., in [31, 32]): theoretical bounds determined by the existent methods of analysis are unachievable for real systems. As a result, modern CSCs transmit signals with the necessary performance but nobody can assess the efficiency of their energy, spectral, and other resources utilization. The listed reasons, including scenario-dependent performance, make the development of the perfect CSCs optimally and fully utilizing their energy and spectral resources an unrealistic task.

In the next section, we show that modulation and feedback resolve the listed problems and enable elaboration of the FCS transmitting signals perfectly, as well as permit to improve their performance criteria.

3. Perfect FCS: transmission using feedback and adaptive modulation

3.1 General principles of FCS transmission

The novelty of the topic makes us begin by considering sufficiently general but not complex systems to simplify the explanation of the main ideas, mathematical tools, methods, and results. Below we consider point-to-point FCS (block diagram in **Figure 1**) assuming that the input signals are Gaussian and channel noises are AWGN, and high quality feedback channel delivers signals from the BSt to the FT with negligibly small errors. One may add that this block diagram, with different formulations of the tasks, was the subject of both early and later research in this field. The material below does not repeat any of these works but summarizes and clarifies their main ideas, approaches to problem-solving, and results to elucidate the difficulties, which had blocked the development of the theory. We also hope that the reader might appreciate the beauty of these works, which came so close to success, but which are now almost forgotten.

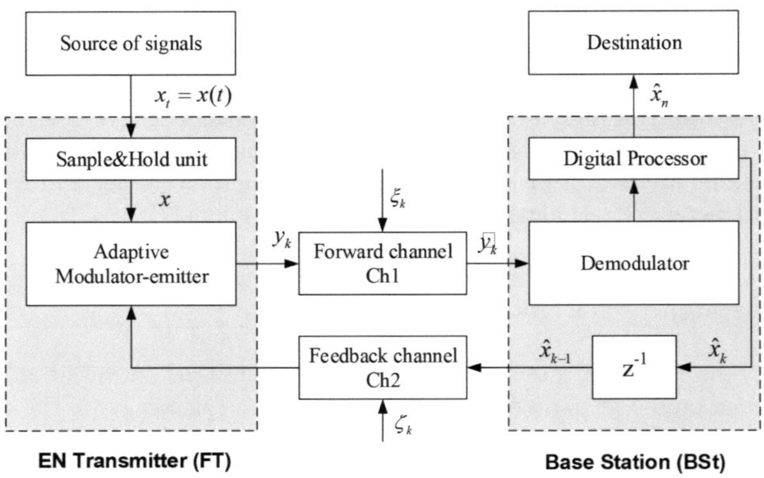

Figure 1.
General block diagrams of point-to-point AFCS.

The analysis that follows is carried out in discrete time. Samples $x^{(m)}$, $m = 1, 2, \ldots$ of Gaussian input signal x_t are transmitted iteratively, each in n cycles, and independently from previous samples. This permits to reduce the analysis of FCS functioning to transmission of a single sample x omitting the upper indices "m." The mean value x_0 and variance σ_0^2 of the samples are assumed to be known. We also assume that the feedback channel Ch2 is realized on high-quality digital components, and that influence of the channel noise on feedback transmission can be neglected. The physical forward channel Ch1 is stationary, memoryless, and its noise ξ_k, $k = 1, \ldots, n$, is AWGN with a double-side spectral power density $N_\xi/2$.

Values x of the transmitted sample are held at the input of the sample and hold unit (S&H) during the time $T_n = n\Delta t_0$ sufficient for the transmission of the sample in n cycles (T_n does not exceed the sampling period $T = 1/2F$; Δt_0 is the duration of the cycle and determines the minimal bandwidth $2F_0 = 1/\Delta t_0$ of the forward and feedback channels; F is the width of the signal baseband). It is assumed that both the FT and BSt have microcontrollers or other signal processing units synchronizing and controlling the work of the transmitter and receivers of FCS.

For every k-th cycle of transmission ($k = 1, \ldots, n$), microcontroller of the forward transmitter (FT) forms the residual signal $e_k = x - \hat{x}_{k-1}$ where $\hat{x}_{k-1}$ is the estimate of the sample computed by the processor of the BSt in previous cycle and delivered to the FT over feedback channel Ch2. Signal e_k is routed to the input of digitally controlled amplitude (AM) modulator-emitter. The signals emitted by FT $\tilde{y}_k$ have the form of high-frequency pulses of the same duration Δt_0:

$$y_k = A_0 M_k e_k = A_0 M_k (x - \hat{x}_{k-1}), (\hat{x}_0 = x_0), \tag{6}$$

where high-frequency (RF) components are omitted; A_0 is the amplitude of the carrier signal; and value of the modulation index M_k is set by the corresponding code previously written into the memory of the FT microcontroller, or delivered to the FT from the BSt over the feedback channel. After demodulation, the signal received by the BSt

$$\tilde{y}_k = AM_k e_k + \xi_k, \tag{7}$$

is routed to the processor of the BSt which computes a new estimate $\hat{x}_k$ of the sample according to the Kalman-type equation:

$$\hat{x}_k = \hat{x}_{k-1} + L_k \left[\tilde{y}_k - \mathrm{E}\left(\tilde{y}_k | \tilde{y}_1^{k-1}\right)\right], (\hat{x}_0 = x_0). \tag{8}$$

The variable A in (7) describes the amplitude $A = A_0\gamma/d$ of the received signal, which depends on the distance d between the FT and BSt, and on the channel gain γ dependent on the propagation losses, characteristics of environment, type and gains of antennas, etc. The gains L_k in (8) determine the rate of convergence of estimates $\hat{x}_k$ and their values, like the values M_k, are determined additionally free parameters stored in the memory of the BSt processor. Value $\mathrm{E}\left(\tilde{y}_k | \tilde{y}_1^{k-1}\right)$ in (8) describes the predicted mean value of the signal y_k computed in the processor of the BSt in previous cycle, and $\tilde{y}_1^{k-1} = \left(\tilde{y}_1, \ldots, \tilde{y}_{k-1}\right)$ denotes a sequence of signals received by the base station in previous cycles.

The BSt processor stores estimate $\hat{x}_k$ and sends it to the FT over the feedback channel. It also resets the gain L_k to the value L_{k+1} and prepares the physical receiver of the station to receive a new signal. Reception of the estimate by the FT initializes the next cycle of the sample transmission: microcontroller computes the residual (estimation error) $e_{k+1} = x - \hat{x}_k$, resets the gain M_k to the value M_{k+1}, and

FCS begins a new cycle of transmission. After n cycles, the final estimate $\hat{x}_n$ is routed to the addressee, while the system recovers its initial state and begins transmission of the next sample.

3.2 Principles and particularities of FCS and AFCS optimization

All the results of pioneering and later research in FCS optimization were obtained using *linear* models of FT transmitters, and this section presents the basic idea of these researches, as well techniques and results of FCS optimization for forward transmitters described by the linear model (7).

Models (6)–(8) are not abstract and describe the sequence of transformations of the signal along its transition over the real units and components of FCS, as well the influence of noises and distortions on the final result of transmission. Moreover, each of these models allows calculation of the changes in the statistical characteristics of signals after each subsequent transformation, and considers the most substantial particularities of this process influencing the work of the system.

Apart from the initially known (given) parameters, these models contain free parameters permitting the designers to regulate the work of particular units and improve the performance of the overall system. For the FCS under consideration, these parameters are $M_1, \dots, M_n$ and $L_1, \dots, L_n$. The basic criterion of the transmission quality is the accuracy ("fidelity" in [1]) of recovery of the signal that is MSE of its estimates $P_k = \mathrm{E}\left[(x - \hat{x}_k)^2\right]$.

The optimization of FCS begins from the definition of algorithm permitting to compute, using the received data $\tilde{y}_1^k$, optimal estimates $\hat{x}_k = \hat{x}_k\left(\tilde{y}_{1_1}^k\right)$ minimizing the MSE P_k for each $k = 1, \dots, n$. According to Bayesian estimation theory (see [33, 34]), in the Gaussian case, these are conditional averages $\hat{x}_k = E\left(x|\tilde{y}_1^k\right)$of random values observed in the presence of AWGN. Moreover, residuals $e_k = x - \hat{x}_k$ and values $\tilde{y}_k$ of the received signal have zero mean values and are mutually orthogonal [34]:

$$\mathrm{E}\left(e_k|\tilde{y}_1^{k-1}\right) = \mathrm{E}\left(x|\tilde{y}_1^{k-1}\right) - \hat{x}_{k-1} = 0; \tag{9}$$

$$\mathrm{E}\left(\tilde{y}_k|\tilde{y}_1^{k-1}\right) = AM_k\mathrm{E}\left(e_k|\tilde{y}_1^{k-1}\right) + \mathrm{E}\left(\xi_k|\tilde{y}_1^{k-1}\right) = 0; \tag{10}$$

$$\mathrm{E}(e_k e_m) = P_k\delta_{mk};\;\; \mathrm{E}\left(\tilde{y}_k\tilde{y}_m\right) = \left(\sigma_\xi^2 + A^2M_k^2\right)\delta_{mk}, \tag{11}$$

where $\delta_{mk} = 1$ for $m = k$ and $\delta_{mk} = 0$ for $m \neq k$.

Substitution of (10) into (8) results in algorithm computing optimal Bayesian estimates that takes an extremely simple form:

$$\hat{x}_k = \hat{x}_{k-1} + L_k\tilde{y}_k. \tag{12}$$

The full transmission-reception algorithm (6)–(8) permits to build a mathematical model of transmission process and to derive the following algorithm for calculation of the mean square error (MSE) of estimates formed by FCS in sequential cycles.

$$\begin{aligned} P_k &= \mathrm{E}\left[(x - \hat{x}_k)^2\right] = \mathrm{E}[(1 - AM_kL_k)\,(\hat{x}_{k-1} - x) + L_k\xi_k]^2 = \\ &= (1 - AM_kL_k)^2P_{k-1} + L_k^2\sigma_\xi^2, \end{aligned} \tag{13}$$

where $P_0 = \sigma_0^2$ and $\hat{x}_k = \mathrm{E}(x|\tilde{y}_1^k)$ are optimal Bayesian estimates (in this case Eq. (8) takes an extremely simple form(12)).

Free parameters M_k, L_k in the right-hand side of (13) do not depend explicitly on the previous values of gains $M_{1,\dots,}M_k$, $L_{1,\dots,}L_k$, which allows the following formulation of the optimization task:

For each k = 1,…,n, one should find values of the gains $M_{1,\dots,}M_k$, $L_{1,\dots,}L_k$, which minimize, under additional conditions and constrains, the MSE of transmission (13).

Beginning the first works, the most widely used additional condition was (and still remains) a constraint on the instant or average power of emitted signals W_0^{sign}. It cannot be greater than the power S_0 of the transmitter:

$$E\left[y_k^2\right] \le S_0 \quad \text{or} \quad \frac{1}{n}\sum_{k=1}^{n} E\left[y_k^2\right] \le S_0. \tag{14}$$

Without loss of generality, one may assume that the power of the FT transmitter and the amplitude of emitted signals are connected by the relationship $S_0 = A_0^2$. In this case, substitution of formula (6) into (14) directly gives the expression for the set of permissible values of modulation index M_k:

$$M_k \le \frac{1}{A_0}\sqrt{\frac{S_0}{P_{k-1}}} = \frac{1}{\sqrt{P_{k-1}}}. \tag{15}$$

According to (13), for every M_k satisfying the inequality (15), MSE of transmission depends on the gains L_k, which in turn determine the rate of convergence of estimates $\hat{x}_k$ the input value x. The extremum of MSE in the set of L_k under definite $M_k$not violating condition (15) can be easily found, and the point of extremum is determined by the formula:

$$L_k^{opt} = \frac{AM_kP_{k-1}}{\sigma_\xi^2 + A^2M_k^2P_{k-1}} = \frac{1}{AM_k}\left(1 - \frac{P_k}{P_{k-1}}\right). \tag{16}$$

The substitution of (16) into (13) gives the following recurrence equation for the corresponding minimal values of MSE:

$$P_k = \frac{\sigma_\xi^2 P_{k-1}}{\sigma_\xi^2 + A^2M_k^2P_{k-1}} = \frac{P_{k-1}}{1 + Q_k^2P_{k-1}}, \tag{17}$$

where parameter $Q_k^2 = A^2M_k^2/\sigma_\xi^2$ describes the SNR at the output of the forward channel.

Formula (17) shows that greater values of modulation index M_k decrease the MSE P_k, that is, improve the quality of transmission. However, the increase of the values M_k is limited by condition (15), and the theoretically achievable minimum of MSE is achievable in practice only if these gains M_k are set to the values:

$$M_k^{opt} = \frac{1}{A_0}\sqrt{\frac{S_0}{P_{k-1}^{\min}}} = \frac{1}{\sqrt{P_{k-1}^{\min}}}; \quad M_1^{opt} = \frac{1}{\sigma_0}, \tag{18}$$

where $P_{k-1}^{\min}$ is defined by Eq. (17) with the values M_k set to the values in (18). The result of the replacement determines the theoretically achievable lower boundary of MSE values described by the relationship:

$$P_k^{\min} = \sigma_0^2 \left(1 + Q^2\right)^{-k}; P_0 = \sigma_0^2, \tag{19}$$

where SNR $Q_k^2 = A^2 M_k^2 / \sigma_\xi^2$ at the forward channel output is constant for each cycle of the sample transmission and takes the values:

$$SNR_k = Q_k^2 = \frac{W^{sign}}{\sigma_\xi^2} = \frac{1}{N_\xi F_0} \left(\frac{A_0 \gamma}{d}\right)^2 = Q^2. \tag{20}$$

Claim 1: Relationships (6)–(8) with the parameters M_k, L_k set, for each $k = 1, \ldots, n$, to the values (16) and (18) determine the optimal transmission-reception algorithm, which contains the information permitting us to design optimal FCS transmitting signals with maximal accuracy (minimal MSE), and this boundary is determined by formula (19). Greater accuracy is not feasible.

Moreover, the optimal transmission-reception algorithm permits us to compute the information characteristics of optimal AFCS prior and posterior entropies, as well as the mean amount of information in estimates $\hat{x}_k$ regarding the values of input samples x:

$$H(X) = \frac{1}{2} \log_2 (2\pi e \sigma_0^2); H\left(X | \hat{X}_k\right) = \frac{1}{2} \log_2 (2\pi e P_k); \tag{21}$$

$$I\left(X, \hat{X}_k\right) = H(X) - H\left(X | \hat{X}_k\right) = \frac{1}{2} \log_2 \left(\frac{\sigma_0^2}{P_k}\right). \tag{22}$$

Taking into the account that the amount of information (22) in estimates is achieved in k cycles, that is, during the time $T_k = k\Delta t_0$, formulas (22) and (19) permit us to evaluate the current values of bit rate of the signal transmission:

$$\begin{aligned} R_{\max}^{FCS} &= \frac{I_{\max}\left(X, \hat{X}_k\right)}{k \Delta t_0} = F_0 \log_2 \left(\frac{\sigma_0^2}{P_k^{\min}}\right) = \\ &= F_0 \log_2 (1 + Q^2) = F_0 \log_2 \left(1 + \frac{W^{sign}}{N_\xi F_0}\right) = C[\text{bit/s}]. \end{aligned} \tag{23}$$

Claim 2. Formula (23) is identical to Shannon's formula (1) and determines the limit bit rate of transmission, that is, the capacity of the system. Moreover, attaining the boundary (23) means that spectral and energy efficiencies of the FCS also attain their limit values and are connected through Shannon's relationship (3). Let us stress that, unlike CSC, the presented relationships determine the approach to the optimal FCS design.

Similar results were obtained in [4, 6, 9] and other works. However, regardless of their correctness, neither the above nor earlier obtained algorithms of transmission in any of their versions could be implemented in practice. Analysis of the reasons showed that the main reason was the omission of saturation effects in the FT.

Another, not less critical reason, and not counted in all formulations of the optimization tasks, has been noted in Section 2, that is, the local dependence of the quality of transmission on the scenario of FCS application. The presented above results confirm this fact directly: values of optimal parameters M_k, L_k as well as all other characteristics of optimal FCS depend on the distance between the FT and the BSt, so that and changes in the FCS's position or surrounding violate the perfect mode of transmission.

In the following sections, we show that modulation and feedback may resolve or at least substantially reduce these problems and make the perfect FCS feasible.

3.2.1 Influence of FT saturation and its elimination

Signals $e_k = x - \hat{x}_{k-1}$ at the input of AM modulators repeat the error of the sample estimate formed in BSt in previous cycle, and their variance is equal to the MSE P_k. This allows the modulation gains to be set, in each cycle, to greater and greater values, which increases SNR at the output of the forward channel and provides superfast growth of the accuracy of the estimates unachievable without feedback.

However, externally completely correct additional condition (14) does not count possible saturation of the modulators or emitters, if the signals $e_k = x - \hat{x}_{k-1}$ exceed their linear range, and an adequate model of the transmitter is to have the form (see also **Figure 2**):

$$y_k = A_0 \begin{cases} M_k e_k & \text{if } \hat{M}_k |e_k| \leq 1 \\ \text{sign } (e_k) & \text{if } \hat{M}_k |e_k| > 1 \end{cases}. \tag{24}$$

In real FT, output range $[-A_0, A_0]$ is fixed except in particular cases, and the width of its input range depends on the value of modulation gain M_k. Setting the values M_k and omitting a consideration of statistics of the signals $e_k = x - \hat{x}_{k-1}$ excludes a possibility of considering saturation of the FT, which appears, if the signal $y_k = A_0 M_k e_k$ crosses the boundaries of its output range. It is worth adding that each saturation during the sample transmission distracts its estimate and causes irreversible loss of information about the sample value.

The probability of the first saturation of FT would appear in the k-th cycle and can be easily evaluated: both signals e_k and y_k are zero mean Gaussian values, and their variances are known. The not complex calculations yield the following relationship:

$$\begin{aligned} \Pr_k^{sat} &= \Pr(|y_k| \geq A_0 | \tilde{y}_1^{k-1}) = \Pr(|e_k| \geq 1/M_k | \tilde{y}_1^{k-1}) \\ &= 1 - \frac{1}{\sqrt{2\pi P_{k-1}}} \int_{-1/M_k}^{1/M_k} \exp\left(-\frac{e_k^2}{2P_{k-1}}\right) de_k = 1 - 2\Phi\left(\frac{1}{M_k P_{k-1}}\right), \end{aligned} \tag{25}$$

where $\Phi(x)$ is a tabulated Gaussian integral (5).

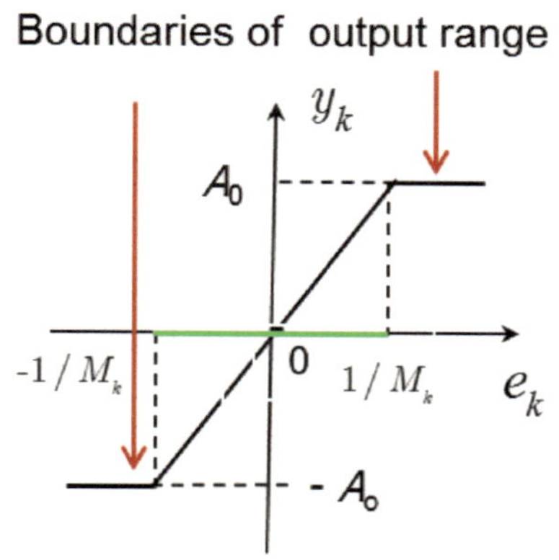

Figure 2.
Static transition characteristic of the transmitter with a finite output range.

Substitution of formula (18) into (25) gives the following evaluation of the probability of the first saturation in k-th cycle of transmission, beginning with the first cycle:

$$\Pr_k^{sat} = 1 - 2\Phi(1) \approx 1 - 0.68 = 0.32,$$

and the probability of its appearance during first five cycles of transmission attains the value $\Pr_{1\div5}^{sat} = 1 - (1 - P^{sat})^5 = 1 - 0.68^5 \approx 0.85$ and quickly tends to unity in next cycles.

One should add that MSE of estimates is weakly sensitive to sufficiently rare cases of FT saturation. However, taking into account that each instance of saturation causes a loss of the sample and $I(X, \hat{X}_n)$ bits of information, the probability of saturation determines the mean percent of erroneous bits in binary sequences delivered to the addressee. These losses can be considered as the BER of transmission (rather bit word error rate—WER but numerically these values are equal). Value of the BER is one of the key characteristics of CSC, and the general tendency in modern communications is to decrease its values to $10^{-6} \div 10^{-8}$ and lesser. Therefore, the setting of modulation index requires closest attention. Investigations showed that the severity of this problem can be sufficiently reduced by employing the feedback channel.

Claim 3. The linear models of the transmitters that are commonly used in formulations of optimization tasks and constraint on the mean power of emitted signals describe the work of transmitters inadequately. This has been and still remains one of the main reasons why FCSs have not been implemented in practice. The solution of this problem allowed for the application of the "statistical fitting condition" [35].

3.2.2 Statistical fitting condition

The previous section and formula (25) show that saturation of the FT can be almost eliminated if the gains M_k are set to the values that guarantee the probability of saturation not greater than a given small $\mu < < 1$, (e.g., $\mu \sim 10^{-4} \div 10^{-8}$) for every sequence $\tilde{y}_1^{k-1}$:

$$\Pr_k^{sat} = \Pr\left(|y_k| \geq A_0 | \tilde{y}_1^{k-1}\right) = 1 - \int_{-1/M_k}^{1/M_k} p\left(e_k | \tilde{y}_1^{k-1}\right) de_k == 1 - 2\Phi\left(\frac{1}{M_k P_{k-1}}\right) \leq \mu. \quad (26)$$

Under fulfilled condition (25), the probability of the sample saturation has the value $1 - (1 - \ \mu)^n \approx n\mu < < 1$. This means almost always (at the confidence level $1 - n\mu$), the FCS would work in optimal mode without saturations, and the percent of erroneous bits would not be greater than μ.

Relationship (26) determines the set of permissible values of the gains M_k and its upper boundary. These values make (26) the equation which, after replacement of the values $1/M_k P_{k-1}$ by the variable α, gives the following relationships for the permissible maximal (optimal) values of modulation gains:

$$M_k^{opt} = \frac{1}{\alpha\sqrt{P_{k-1}^{\min}}}; (k = 1, .., n), \quad (27)$$

where parameter a satisfies the equation:

$$\Phi(\alpha) = \frac{1}{\sqrt{2\pi}} \int_0^{\alpha} \exp\left(-\frac{z^2}{2}\right) dz = \frac{1-\mu}{2}. \tag{28}$$

Claim 4. Setting the modulation index to the values (27) prevents the appearance of saturation and the FCS with non-linear FT transmits the signals almost always in linear mode.

Claim 5. Setting the gains M_k to the values (27) not only removes saturation but also:

a. reduces α times the amplitude of emitted signals: $A_0 \to A_0/\alpha$,

b. reduces α^2 times the power of emitted signals and SNR at the output of the channel:

$$W^{sign} = \frac{A^2}{\alpha^2}; SNR_k = Q_k^2 = \frac{W^{sign}}{\sigma_\xi^2} = \frac{1}{N_\xi F_0} \left(\frac{\gamma A_0}{\alpha d}\right)^2 = Q^2. \tag{29}$$

The structure and form of the basic relationships for the MSE, bit rate, and effectiveness remain the same. Changes in Shannon's formula (23) for the capacity and in the other relationships affect only the values of amplitude and power of emitted signals (29).

Claim 6. Reduction of the power of the emitted signal decreases the SNR and capacity of the system but makes the perfect FCS feasible, and these systems transmit signals with limit energy-spectral efficiency optimally, as well as completely utilizing the frequency and energy resources of the FT. Moreover, the absence of coders reduces their energy consumption.

To distinguish the considered systems and the FCS, and to focus on the new systems, in what follows, we use a new abbreviation to refer to these systems: AFCS (adaptive FCS).

Let us remind that these relationships were derived under assumption that the feedback channels are ideal. This is not an unrealistic suggestion: relatively inexpensive modern CSCs provide virtually error-free short-range transmission. If necessary, the quality of the channel can also be improved by increasing the power of transmitters—the BSt have sufficiently large, if not unlimited energy resource.

We should add that the discussed relationships are a particular case of the more general optimal transmission-reception algorithm for the statistically fitted AFCS with noisy feedback [36–39], see also **Table 1** below. It is worth stressing that these algorithms have the same structure and form as those presented in pervious sections. The only but principle difference concerns the expression for the MSE of transmission, which takes the form:

$$P_k^{\min} = \frac{\left(\sigma_\xi^2 + A^2 M_k^2 \sigma_\nu^2\right) P_{k-1}^{\min}}{\sigma_\xi^2 + A^2 M_k^2 \left(\sigma_\nu^2 + P_{k-1}^{\min}\right)} = \left(1 + Q^2\right)^{-1} \left[1 + Q^2 \frac{\sigma_\nu^2}{\left(\sigma_\nu^2 + P_{k-1}^{\min}\right)}\right] P_{k-1}^{\min}, \tag{30}$$

where variable σ_ν^2 is the variance of the errors ν_k in the signals (controls) $\hat{x}_k$ transmitted from the BSt to FT over the feedback channel with AWGN. Under $\sigma_\nu^2 = 0$, this formula coincides with (17). Analysis of formula (30) gives comprehensive answers to many questions concerning the behavior of feedback systems.

	Algorithm	Parameters
Initial values	$\hat{x}_0 = x_0$	$P_0 = \sigma_0^2$
Signal at the modulator input	$e_k = x - \hat{x}_{k-1} + v_k$	$\alpha : \frac{1}{\sqrt{2\pi}} \int_0^{\alpha} \exp\left(-\frac{z^2}{2}\right) dz = \frac{1-\mu}{2}$
Emitted signal	$y_k = A_0 M_k e_k$	$M_0 = \frac{1}{\alpha\sigma_0}; M_{k/k\geq 1} = \frac{1}{\alpha\sqrt{\sigma_\nu^2 + P_{k-1}}}$
Received signal	$\tilde{y}_k = A y_k + \xi_k$	$A = \frac{\gamma}{d} A_0$
Estimate computing	$\hat{x}_k = \hat{x}_{k-1} + L_k \tilde{y}_k$	$L_k = \frac{A M_k P_{k-1}}{\sigma_\xi^2 + A^2 M_k^2 P_{k-1}} = \frac{1}{A M_k}\left(1 - \frac{P_k}{P_{k-1}}\right)$
Basic equation for MSE	$P_k = \left(1 + Q^2\right)^{-1}\left[1 + Q^2 \frac{\sigma_\nu^2}{\left(\sigma_\nu^2 + P_{k-1}\right)}\right] P_{k-1}; k = 1, \ldots, n$	

Table 1.
Basic relationships for modeling and design of optimal AFCS with non-ideal feedback channel ($\sigma_\nu^2>0$; index "opt" in the parameters M_k^{opt}, L_k^{opt} is omitted).

The main result is the confirmation that the capacity of FCS and AFCS does not depend on the feedback noise in the initial interval of the sample transmission, and is determined by Shannon's formula (1). However, since the moment n^*, when the MSE of estimates attains the values of σ_ν^2 order, the capacity of AFCS begins to decrease and monotonically tends to zero ([36] and later works).

A summary of the relationships sufficient for the development of a MATLAB model of optimal AFCS and simulation experiments is presented in **Table 1**. Moreover, these seemingly simple relationships were used to design a prototype (demonstrator) of the perfect AFCS, discussed in the next section.

If a relative error of the sample transmission $\delta = \Delta/\sigma_0$ is to be attained in minimal time, and feedback is ideal, absolute error Δ of the final estimates should be not greater than σ_ν, and the corresponding minimal time of transmission is of the order:

$$T_\delta = \frac{n_\Delta}{2F_0} = -\frac{1}{2F_0 \log_2\left(1 + Q^2\right)} \log_2\left(\frac{\Delta^2}{\sigma_0^2}\right) = -\frac{\log_2 \delta}{C^{Ch1}}. \qquad (31)$$

where n_Δ determines the necessary number of cycles (see e.g. formula (27) in [37] or (22) in [38]). Formula (31) also determines the baseband $F \leq F_\delta = C^{ch1}/2\log_2\delta$ of the signals that can be transmitted by the AFCS at maximum rate, with the given accuracy and limit energy-spectral efficiency.

3.2.3 Particular role of MSE criterion

As noted above, the basic criterion for transmission quality is the accuracy (in [1] "fidelity") of the signals' recovery. For the CS transmitting analog signals, this is the MSE of their estimates. The importance of MSE is due to several factors. First of all, for arbitrary linear channels with AWGN, which transmit Gaussian signals, MSE P_k determines the amount of information determined by following general relationship ([32, 34], see also (22)):

$$I\left(X, \hat{X}_n\right) = \frac{1}{2} \log_2\left(\frac{\sigma_0^2}{P_n}\right) \text{ [bit/sample]}. \qquad (32)$$

So, if CSs transmit the samples each in n cycles, that is, during $T_n = n\Delta t_0 = n/2F_0$ [s], the final estimates $\hat{x}_k$ deliver to addressee the amount of information with the bit rate

$$R_n = \frac{I(X, \hat{X}_n)}{n\Delta t_0} = F_0 \log_2\left(\frac{\sigma_0^2}{P_n}\right) \text{ [bit/s]}, \tag{33}$$

independent of whether the system is optimal or not. This value determines the spectral efficiency of the sample transmission R_n/F_0 that is at the AFCS output.

The iterative principle of transmission permits us to introduce the measure more informative than (33): the *instant* bit rate, determined by the following relationship:

$$\Delta R_k = \frac{I(X, \hat{X}_k) - I(X, \hat{X}_{k-1})}{\Delta t_0} = F_0 \log_2 \frac{P_{k-1}}{P_k} \text{ [bit/s]}, \tag{34}$$

which describes the increment of information in sequentially computed estimates $\hat{x}_k$, $(k = 1, \dots, n)$. In turn, formulas (33) and (34) define the final and instant spectral efficiencies of transmission, respectively:

$$\frac{R_n}{F_0} = \frac{1}{n} \log_2 \frac{\sigma_0^2}{P_n}; \; \frac{\Delta R_k}{F_0} = \log_2 \frac{P_{k-1}}{P_k}. \tag{35}$$

The general expression for the energy efficiency of transmission can be defined as follows:

$$\frac{E_n^{bit\ \mathrm{AFCS}}}{N_\xi} = \frac{W_n^{sign}}{N_\xi R_n} = \frac{n W_n^{sign}}{N_\xi F_0 I(X, \hat{X}_n)} = \frac{nQ^2}{\log_2 \frac{\sigma_0^2}{P_n}}, \tag{36}$$

which shows that, unlike spectral efficiency, this characteristic of the CS performance depends not only on the MSE, but also on the SNR Q^2 at the forward channel output, which requires additional measurement.

Another particularity of the MSE, which is not currently utilized in communications, is its analytical formulations have empirical analogs, as well as well-studied and widely used methods of their evaluation. In our research, the following method is used. However, as it follows from (36), evaluation of the energy efficiency, in the general case, requires additional measurement of SNR Q^2. In our research, the following method is used.

The FT generates and sends to the BSt a testing sequence of M random Gaussian samples $(x^{(1)}, \dots, x^{(M)})$ each generated with the same mean value x_0 and variance σ_0^2 using corresponding codes written into memory units of microcontrollers of the FT transmitter and processor of the BSt (or generated by PC). The BSt processes the received signals, computes optimal estimates $\left(\hat{x}_1^{(1)}, \hat{x}_2^{(1)} \dots, \hat{x}_{n-1}^{(M)}, \hat{x}_n^{(M)}\right)$ of input samples which, and stored values $(x^{(1)}, \dots, x^{(M)})$, allow to compute empirical values of the MSE P_k using the known relationship:

$$\hat{P}_k = \frac{1}{M} \sum_{m=1}^{M} \left[x^{(m)} - \hat{x}_k^{(m)}\right]^2, (k = 1, \dots, n) \tag{37}$$

next used for the evaluation of the bit rate (35) and energy-spectral efficiency of AFCS. In practice, it is more convenient to compute these values using the MSE expressed in dB:

$$\mathrm{MSE}_k[\mathrm{dB}] = 10 \log_{10}\left(\frac{\hat{P}_k}{\sigma_0^2}\right) \text{ [dB]}. \tag{38}$$

as well as normalized root square (relative error of transmission) $\delta = \sqrt{P_k}/\sigma_0$.

3.3 Adaptive auto-adjusting AFCS to the scenario of application

The adjusting algorithm uses the "resonance" effect that is increase of MSE, if the values of parameters L_k, M_k decline from their optimal values (16), (27). The effect is illustrated in **Figure 3** which shows the changes of relative errors of transmission $\delta = \sqrt{P_k}/\sigma_0$ (normalized root mean square error—RMS) under gains M_k set to the values (27) and gain $L_k^* = L_k(1 + \delta_L)$, where L_k has the value (16) and δ_L is a variable parameter.

To adjust the parameters, the system utilizes two identical testing sequences of Gaussian samples $\left(x^{(1)}, \ldots, x^{(M)}\right)$ whose codes are stored in the memory of the BSt processor and the FT microcontroller (or BSt transmits these sequences to FT over feedback). All the samples in the testing sequence are zero mean Gaussian values with known variance σ_0^2.

In the first cycle, the modulation index M_1^{opt} is set to the known value $1/\alpha\sigma_0$, and the FT sends to the BSt the written testing sequence of samples $\left(x^{(1)}, \ldots, x^{(M)}\right)$, the same one that is stored in the processor of BSt. The BSt processor computes the estimates $\left(\hat{x}_1^{(1)}, \ldots, \hat{x}_1^{(M)}\right)$ and values of MSE P_1. It also searches for the minimizing MSE value $\hat{L}_1^{opt}$ and computes the corresponding value $\hat{P}_1^{\min}$. The computed values $\hat{L}_1^{opt}, \hat{P}_1^{\min}$, and $\left(x^{(1)}, \ldots, x^{(M)}\right)$ are stored. Simultaneously, the BSt sends values $P_1^{\min}$ and estimates $\left(\hat{x}_1^{(1)}, \ldots, \hat{x}_1^{(M)}\right)$ to the FT. Reception of these data initiates the second cycle of AFCS adjusting:

In this cycle, the microcontroller of the FT forms the sequence of signals $e_2^{(m)} = x - \hat{x}_1^{(m)}$, $(m = 1, \ldots, \mathrm{M})$, computes the optimal value of the gain $M_2^{opt} = 1/\alpha\sqrt{P_1^{\min}}$, and sets the gain of AM modulator to this value. So adjusted, the FT transmits the sequence of signals $y_2^{(m)} = A_0 M_2^{opt} e_2^{(m)}$ to the BSt, which processes the received sequence in the same way as in first cycle. The computed values $P_2^{\min}, L_2^{opt}$ are stored, sequence $\left(\hat{x}_1^{(1)}, \ldots, \hat{x}_1^{(M)}\right)$ is replaced by the new sequence $\left(\hat{x}_2^{(1)}, \ldots, \hat{x}_2^{(M)}\right)$, and together with $P_2^{\min}$ is transmitted to the FT. The receipt of these values initializes the next cycle of adjusting realized according to the same scheme as in the previous cycle. The subsequent cycles repeat these operations. After the M-th cycle, the adjusted AFCS begins nominal functioning.

The duration and frequency of adjustments depend on the dynamics of scenario changes, processors' rate, channel bandwidth, requirements for the accuracy of estimates, environmental characteristics, and other factors.

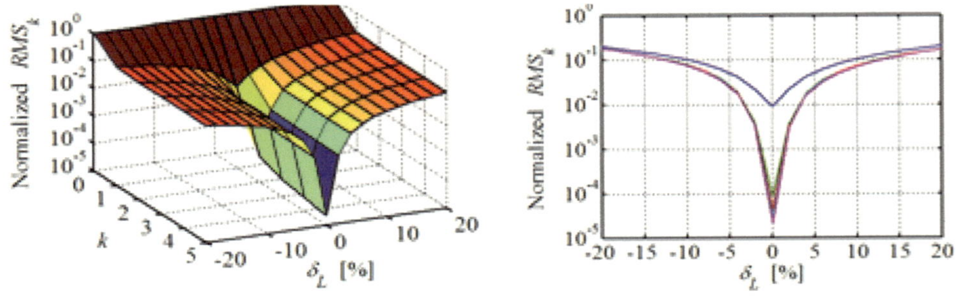

Figure 3.
Changes of mean relative error of estimates depending on the deviation of the gains L_k *from optimal values* L_k^{opt} *under fixed optimal gains* M_1^{opt}*.*

4. Experimental study of AFCS functioning

The prototype of AFCS was designed on the basis of the optimal transmission-reception algorithm(6)–(8), using and parameters set to the values (16), (18), (or in **Table 1**, for $\sigma_v^2 = 0$), and general principles of AFCS transmission described in Section 3.1. The layouts of the transmitting (FT) and receiving (BSt) modules are shown in **Figure 4a**, **b**.

The transmitter was realized using narrowband adaptive AM modulator followed by the programmable voltage-controlled oscillator VG7050EAN (power 10 dBm, carrier frequency 433.2 MHz). The feedback channel was realized using digital receiver RFM31B-S2 and transmitter RFM23B (power 27 dBm, carrier frequency 868.3 MHz). This ensured virtually ideal feedback transmission of signals in the indoor and outdoor experiments carried out at distances to 100 meters (straight line view, FT with ceramic mini-antennas, BSt with quarter-wave antennas).

At the beginning of every new series of experiments, a self-adjusting algorithm was activated, which set the parameters M_k and L_k to the values optimal for the given scenario.

The main measured characteristic of the prototype was the *dependence of* MSE_n[dB] *on the number of transmission cycles. The experiments were* carried out at different distances between the FT and BSt. Typical dependencies of MSE_n[dB] on n at the distances of 40, 50, and 75 meters are shown in **Figure 5**.

The plots are presented in the decibel scale, and the nearly linear dependence of the measured values MSE_n[dB] on the number of cycles means that, on a linear scale, MSE decreases exponentially. According to the results of Section 3.2 (formula (19)), this is possible, if the system transmits signals perfectly, with a bit rate *equal to the capacity* of the system. In this case, spectral and energy efficiencies of transmission also attain the limit values.

Moreover, plots in **Figure 5** allow for a sufficiently accurate evaluation of the characteristics of the system. With this aim, let us rewrite the expression of spectral efficiency (33) in the decibel scale in the form (the confirmed close to perfect transmission permits us to write that $R = C$):

$$\frac{C_n}{F_0} = \frac{3.32}{n} \log_{10} \frac{\sigma_0^2}{P_n} = -\frac{0.332}{n} \text{MSE}_n[\text{dB}]. \tag{39}$$

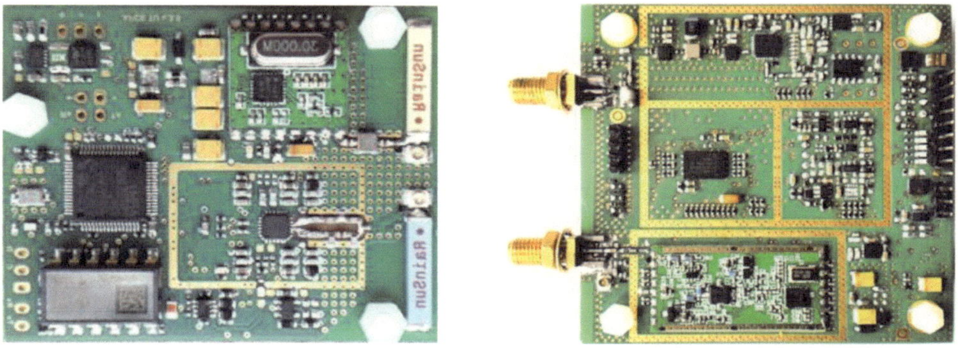

(a) (b)

Figure 4.
Layout of PCB modules of prototype of perfect AFCS: (a) forward transmitter integrated with sensor; (b) base station.

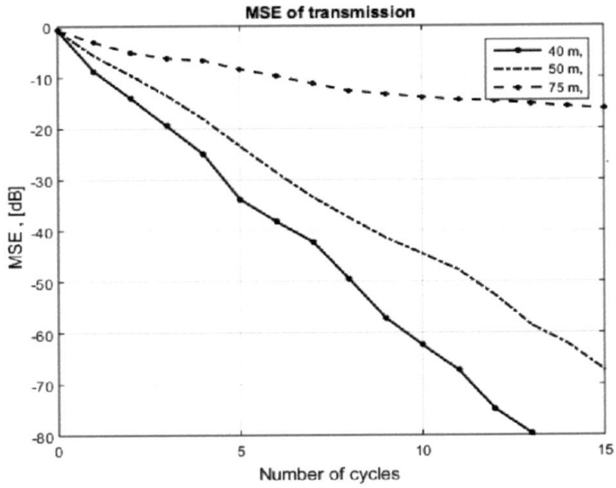

Figure 5.
Values $\mathrm{MSE_k[dB]} = \mathrm{f(k)}$ *measured at the distances of 40, 50, and 75 meters (indoor experiments).*

Distance (m)	40	50	75
$MSEn$ [dB]/n	−6.25	−4.4	−1.4
C^{syst}/F_0	2.07	1.46	0.465
Q^2	3.21	1.75	0.375
$E_n^{bit\ syst}/N_\xi$	1.76	1.19	0.8
$E_n^{bit\ syst}/N_\xi$ [dB]	1.85	0.76	−0.92

Table 2.
Measured characteristics of the prototype.

The linear approximation of the plots in **Figure 5** in first $n = 10$ cycles allows for evaluation of the ratio $\mathrm{MSE}_n[\mathrm{dB}]/n$ directly from the plots. Substitution of these values into (39) gives the numerical estimates of the spectral efficiency for each distance. In turn, perfect mode of transmission allows for the evaluation of the power efficiency using formula (3) and measured values of the spectral efficiency C_n/F_0. The obtained results are presented in **Table 2** and illustrated on the energy-spectral plane in **Figure 6**.

The plots in **Figure 7a, b** illustrate the results of measurements carried out at the fixed distance sequentially, with the time interval in 1–3 minutes.

All the experiments confirmed the existence of the initial interval $1 \le k \le n^*$ of the perfect transmission, as well as the efficiency of the developed algorithm of AFCS auto-adjusting. On average, at the distances of 40–50 m, the accuracy of transmission attained the values of 10^{-4} order in 5÷7 cycles of the sample transmission. In several experiments, accuracy attained the order of $10^{-7} \div 10^{-10}$ and greater, but the results were non-stable (see, e.g., plots in **Figure 7a, b**).

It was also noted the growing influence of the external disturbances and noises on the further changes of MSE if it attained sufficiently small values. Since this time, MSE decreased with the growing fluctuations, sometimes regularly but at the smaller rate. This could not be an effect caused by saturations: experiments were carried out under saturation factor $\alpha = 3$ that reduced the cases of saturations to 0.5–1 percent of the transmitted samples. A study of the AFCS functioning in real

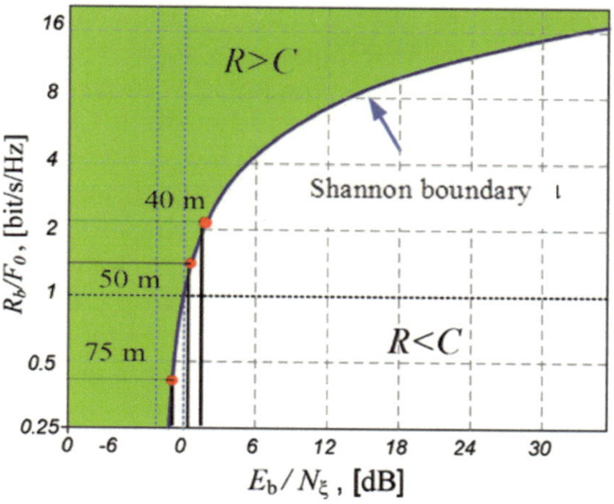

Figure 6.
Two-dimensional plot presenting values (red points) of the energy-spectral efficiency of optimal AFCS measured at the distances in 40, 50, 75 meters.

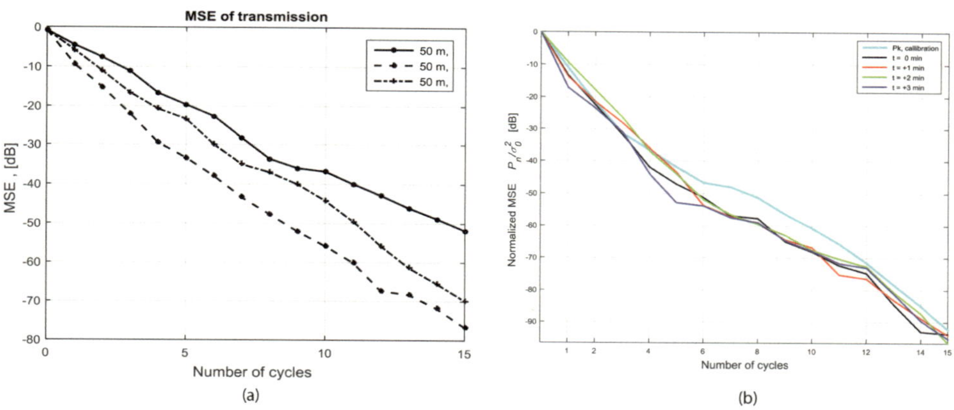

Figure 7.
Dependencies $MSE_k[dB] = f(k)$ *measured with the intervals in 2–3 minutes at the distance of 50 m (indoor experiments): (a) power of FT transmitter 10 mW, ceramic mini-antennas; (b) power of FT transmitter 1 mW, quarter-wave antennas.*

conditions is to be continued. It is worth noting that the utilization of the quarter-wave antennas provided sufficiently stable sample transmission at distances upto 80–90 m using FT with the power reduced to 1 mW.

The experiments confirmed the feasibility of the perfect AFCS, as well as the capability of modulation and feedback to ensure the perfect signal transmission.

5. Discussion of results

The chapter gives a brief outlook of the approaches to design of the currently not used class of communication systems which may substantially improve the efficiency and performance of the wireless low-power transmission. The presented results develop excellent but not finished and today almost forgotten research in FCS theory carried out in the years 1960–1970. These investigations were first steps toward the formation of the second direction in information theory: the theory of the systems with feedback channels. However, in the middle of 70s, the research was hampered.

The main reason for the difficulties was the lack of practical results. Another, less obvious source of failures was the omission of possible saturation of modulators or emitters in the forward transmitters. The not less crucial obstacle also was not discussed in the literature which is the dependence of the CS performance on the scenarios of application.

The chapter shows how modulation and feedback permit to resolve these difficulties. The results of research confirm the general conclusion of the pioneering research: FCS may transmit signals without coding perfectly, that is in real time, with a bit rate equal to the capacity of systems, and with the limit accuracy of the signals recovery. Moreover, the only difference between the relationships presented in the chapter and those presented in earlier works is the numerical values of modulation index.

The rapidly developing wireless networks (WN) utilize a great number of short-range low-energy end-node (EN) transmitters. Nowadays, all of them employ the coding and advanced digital technical solutions. The level and number of requirements for EN transmitters grow permanently. However, as noted by many authors, possibilities to improve the performance of the PHY layer of WN are almost exhausted. We discussed this in the beginning of the chapter. Moreover, all networks utilize, on the mass scale, feedback channels.

What can we conclude? Despite its great merits, coding is losing its advantages and is being used in low-power EN transmitters designed for the short-range transmission ("one mile zone" and shorter). The main task of these transmitters is the reliable and secure delivery of relatively small amounts of information to the BSt or master node, if possible, with minimal delay. The great rates (except for rare exclusions) are not necessary: this is the task for BSt which communicate with other BSt or higher level stations.

Today's problems of the EN transmitters design are prosaic: they should be as less energy-consuming as possible to increase the duration of continuous work ("lifetime") and to reduce the requirements of the energy sources. They should be resistant to inter-channel disturbances caused by nearby EN, as well as should have minimal complexity to decrease the production and deployment costs. It is also desirable that they have low emission, small size, light units, etc.

Reduction of the energy consumption inevitably causes the reduction of the power of the EN transmitters and that crucially decreases the quality and range of transmission. Compensation of the losses requires application of more efficient correcting codes and more complex coders, as well as the extension of the channel bandwidth. In turn, wide band transmission creates sufficiently powerful inter-channel disturbances in closely placed EN. The result is the appearance of complex technical solutions suppressing these distortions or, vice versa, utilizing them for improvement of the signals recovery. The list of tradeoffs between different requirements of the systems is large, and coding has no efficient answers to these questions.

Having no alternatives, the industry has no other choice but to transfer known principles of long-distance CS with coding (CSC) design and technologies to the design of low-power EN transmitters produced on the scale by the orders greater than powerful CSC transmitters. The not too essential for long-distance communications, constraints on the power of the transmitters, requirements for the bandwidth and other constrains became the crucial considerations in the wireless EN design. From our point of view, the greatest stumbling block is that CSCs do not allow for a development of the systematic approach to their optimization similar to the Bayesian approach described in the chapter. Codes have no parameters permitting their adjustment to the changes of characteristics of the channel, nor allow the formulation of mathematical models accounting for all the transformations of signals as they pass through the transmitter, channel, and receiver. AFCSs do have such possibilities.

Moreover, the signals generated by the transmitters of CSC are discrete, their form is fixed and in no way depends on the input signals. Information is delivered by combinations of the symbols of code. There is no possibility to regulate the quality of the transmission aside from the external regulation of the power of transmitter or switching the codes. Meanwhile, the quality of transmission provided by the adjusted perfect AFCS depends on the scenario of their application, but always attains the limit or close to the limit values.

General evaluation of the future perspectives of AFCS: currently, almost all CS and networks have feedback channels, and AFCS could solve many of the aforementioned and other problems. Below, we attach a summary of possibilities of the perfect AFCS, which have been established and verified in [36–41] and other works.

1. Perfect AFCSs provide the most energy-spectral efficient transmission of signals in real time with the limit energy-spectral efficiency, bit rate equal to the capacity of forward channel, and minimal MSE of the signals reception.

2. The absence of coders allows for the construction of a full mathematical model of transmission, from the source of signals to the BSt processor. This model enables the formulation of a clear analytical criterion (MSE), the application of Bayesian estimation theory, and the derivation of optimal transmission-reception algorithms determining the approach to the perfect AFCS design.

3. Feedback channel and optimal transmission algorithms enable the development of adaptive algorithms adjusting the parameters of AFCS to the environment changes. This permits the system to maintain the perfect mode of transmission in different, also non-stationary scenarios. De facto, the system regulates its own capacity, adjusting it to the changes of environment.

4. The side effect of AFCS adjustment is that the BSt computes the on-line estimates of MSE, SNR, and capacity of the system. These data permit to evaluate the current energy-spectral efficiency of transmission and to decrease the losses of energy regulating the number of cycles maintaining the required accuracy of the signal recovered.

5. The analytical expression for MSE of transmission has an empirical analog, whose values can be measured and used for evaluation of the performance of every CS used for the analog signals transmission. As shown in Section 3, MSE permits us to determine the quality of transmission, bit rate, as well spectral efficiency of every CS. For the perfect AFCS, minimal MSE determines the energy efficiency of transmission.

6. Signals emitted by the FT have the form of realizations of the (stationary) pulse white Gaussian noise. The amplitudes of each emitted pulse depend on random values of the signals e_k at the input of the modulator, and values of the adjusted parameters M_k depend on the scenario, which excludes data interception and ensures secure forward transmission. The full protection of the system depends on protecting the feedback channel and can be provided by applying well-protected codes and directed antennas.

7. The absence of coding units simplifies the architecture of the FT, as well as reduces their energy consumption, complexity, and cost, which allow for the development of efficient battery-less AFCS.

8. The FT transmitters can be realized in analog, digital, and mixed technologies. The results of analysis show that the most preferable form of realization would be the software implementation of the FT. Optimal transmission-reception and adjusting algorithms contain all basic information for the development and implementation of the software (SDR) version of the FT. Moreover, this software can be used for the reconfiguration of transmitters and the extension of possibilities of their utilization and functional possibilities of the EN as a whole.

9. Preliminary research [42] showed that perfect AFCS can be also used, virtually without modification of transmission scheme and algorithms, for the transmission of short codes. These codes can be parts of the longer code routed to the input of the AFCS or formed by digital sensors. The set of the codes can be converted into the uniform set of analog values, and high accuracy of AFCS transmission will ensure reliable resolution of the received signals.

10. The designed prototype of AFCS is the first "living" example of the system that transmits signals without coding perfectly, and this was confirmed by the results of experiments.

Most of the listed capabilities of AFCS are not feasible for the CS with coding.

In this chapter, we considered only scalar (point-to-point) AFCS which employ the AM transmission, but the theory allows for the extension to optimization of the multi-channel FCS. Moreover, AM is only one of three types of modulation, and each has its own limited operating range. It would be important to investigate the systems with the FM and PM modulations—this could give a new classes of perfect FCS transmitting signals without abnormal errors and with the limit energy-spectral efficiency. It is also worth noting that statistical fitting condition (26) can be used for the optimization of different classes of estimation, controlling, measurement, and signal processing systems.

In conclusion, let us repeat the questions asked in [19]: "Is the PHY layer dead? ... whether the research directions taken in the past have always been the right choice and how lessons learned could influence future policy decisions?"

Acknowledgements

The author expresses a deep gratitude to the young colleagues Ievgen Zaitsev, Borys Jeleński, and Jan Piekarski, as well as to the colleague Henryk Chaciñski for their active participation in the elaboration of the AFCS prototype and creative engineering thinking.

Acronyms

CS	communication system
CSC	communication system with coding
FCS	feedback communication system
AFCS	adaptive feedback communication system
WN	wireless network
EN	end node
FT	forward transmitter

BSt	base station
PHY	physical layer of network
AWGN	additive white Gaussian noise
SNR	signal-to-noise ratio
BER	bit error rate
MSE	mean square error
MMSE	minimal mean square error
RMS	root mean square

Author details

Anatoliy Platonov
Institute of Electronic Systems Warsaw, Warsaw University of Technology, Poland

*Address all correspondence to: plat@ise.pw.edu.pl; platon945@gmail.com

References

[1] Shannon CE. A mathematical theory of communications. Bell System Technical Journal. 1948;**27**(1):379-423 623-656

[2] Shannon CE. Communication in the presence of noise. Proceedings of the IRE. 1949;**37**(1):10-21

[3] Elias P. Channel capacity without coding. In: Baghdady EJ, editor. MIT Research Laboratory of Electronics, Quarterly Progress Report 1956: 90–93, Reprinted in Lectures on Communication System Theory. New York: Mc Grow Hill; 1961

[4] Omura JK. Signal optimization for channels with feedback [dissertation]. Stanford Electronics Lab; 1962. Stanford, Rept. SEL-66-068

[5] Goblick TJ. Theoretical limitations on the transmssion of data from analog sources. IEEE Transactions on Information Theory. 1965;**1**(4):558-567

[6] Kailath T. An application of Shannon's rate-distortion theory to analog communication over feedback channels. Proceedings of the IEEE. 1967; **55**(6):1102-1103

[7] Schalkwijk JPM, Kailath T. A coding scheme for additive noise channels with feedback - part I. IEEE Transactions on Information Theory. 1966;**12**(2):172-182

[8] Schalkwijk JPM. A coding scheme for additive noise channels with feedback - part I: Bandlimited signals. IEEE Transactions on Information Theory. 1966;**12**(2):183-188

[9] Schalkwijk JPM, Bluestein LI. Transmission of analog waveforms through channels with feedback. IEEE Transactions on Information Theory. 1966;**13**(4):617-619

[10] Omura JK. Optimal transmission of analog data for channels with feedback. IEEE Transactions on Information Theory. 1968;**14**(1):38-43

[11] Butman S. A general formulation of linear feedback communication systems with solutions. IEEE Transactions on Information Theory. 1969;**15**(3):392-400

[12] Butman S. Rate distortion over band-limited feedback channels. IEEE Transactions on Information Theory. 1971;**16**(1):110-112

[13] Fang R. Unification of linear information feedback schemes for additive white Gaussian noise channels. IEEE Transactions on Information Theory. 1970;**16**(11):786-789

[14] Lucky RW. A survey of the communication theory literature: 1968-1973 [invited paper]. IEEE Transactions on Information Theory. 1973;**19**(5):25-39

[15] Gallager RG. Variations on a theme by Schalkwijk and Kailath. IEEE Transactions on Information Theory. 2010;**56**(1):6-17

[16] Hunn N. Essentials of Short-Range Wireless. New York: Cambridge University Press; 2010

[17] Jonson H, Graham M. High Speed Signal Propagation: Advanced Black Magic. New Jersey: Prentice Hall; 2003

[18] Costello DJ, Forney GD. Channel coding: the road to channel capacity [survey]. Proceedings Of IEEE. 2007; **95**(6):1150-1178

[19] Dohler M, Heath RW, Lozano A, Papadias CB, Valenzuela RA. Is the PHY layer dead? IEEE Communications Magazine. 2011;**49**(4):159-166

[20] Goldsmith A. The next wave in wireless technology: challenges and

solutions [keynote]. In: IEEE Wireless Communications and Networking Conference (WCNC). Budapest: IEEE; 2009. Avalavle from: http://wcnc2009.ieee-wcnc.org/keynotes.html

[21] The death of 5G?, Thematic Series of e-Papers in IEEE ComSoc Technology News, 2015–2016, Ed. in Chief, Gatherer, A

[22] Goldsmith A. The road ahead for wireless technology: Dreams and challenges [keynote]. In: International Symposium on Wireless Communication and Systems (ISWCS). Poznan; 2016. Available from: http://iswcs2016.radiokomunikacja.edu.pl/welcome/menu/keynotes#k1

[23] Bedrosian E. Spectrum conservation by efficient channel utilization. IEEE Communications Society Magazine. 1997;**15**(7):20-27

[24] Lee JS, Miller LE. CDMA Systems Engineering Handbook. Boston-London: Artech House; 1988

[25] Sklar B. Digital Communications Fundamentals and Applications. New Jersey: Prentice Hall; 2001

[26] Chen Y, Zhang S, Shugong X, Li GY. Fundamental trade-offs on green wireless networks. IEEE Communications Magazine. 2011;**49**(6):30-37

[27] Dobrushin RL, Prelov VV. Information Theory, Encyclopedia of Mathematics. 2011. Avalable from: http://www.encyclopediaofmath.org/index.php?title=Information_theory&oldid=18981

[28] Bogucka H, Conti A. Degrees of freedom for energy savings in practical adaptive wireless systems. IEEE Communications Magazine. 2011;**49**(6):38-45

[29] Rysavy P. Challenges and considerations in defining spectrum efficiency. Proceedings of the IEEE. 2014;**102**(3):386-392

[30] Li GY, Xu Z, Xiong C, Yang C, Zhang S, Chen Y, et al. Energy-efficient wireless communications: Tutorial, survey, and open issues. IEEE Wireless Communications. 2011;**18**(6):28-35

[31] Roman S. Coding and Information Theory. Berlin: Springer-Verlag; 1992

[32] Haykin S. Communication Systems. 3rd ed. Chichester: Wiley; 1994

[33] Van Trees HL. Detection, Estimation and Modulation Theory. New York: Wiley; 1972

[34] Balakrishnan AV. Kalman Filtering Theory: Optimization Software. New York: Inc. Publications Division; 1984

[35] Platonov A. Optimal identification of regression-type processes under adaptively controlled observation. IEEE Transactions on Signal Processing. 1994;**42**(9):2280-2291

[36] Płatonov A. Analytical Methods in Analog-Digital Adaptive Estimation Systems Design. Publication Office of Warsaw University of Technology; 2006

[37] Platonov A. Optimization of adaptive communication systems with feedback channels. In: IEEE Wireless Conference on Communications and Networking (WCNC). Budapest: IEEE Xplore; 2009. pp. 93-96

[38] Platonov A. Capacity and power-bandwidth efficiency of wireless adaptive feedback communication systems. IEEE Communications Letters. 2012;**16**(5):573-576

[39] Platonov A. Information theory: two theories in one. Proceedings of SPIE. 2013;**8903**:8903G1-8903G16

[40] Platonov A, Zaitsev I. New approach to improvement and

measurement of the performance of PHY layer links of WSN. IEEE Transactions on Instrumentation and Measurement. 2014;**63**(11):2539-2547

[41] Platonov A, Zaitsev I, Jeleński B, Opalski LJ, Piekarski J, Chaciński H. Prototype of analog feedback communication system: First results of experimental study. In: 4th IEEE International Black Sea Conference on Communications and Networking (BlackSeaCom). Varna, Bulgaria: IEEE Xplore; 2016. pp. 1-3

[42] Platonov A. Limit energy-spectral efficient transmission of signals and short codes using adaptive analog modulators. In: International Conference on Signals and Electronic Systems. Warsaw: IEEE Xplore; 2018. pp. 87-92

Chapter 2

Digital Algebraic Method for Processing Complex Signals for Radio Monitoring Systems

Sergey G. Rassomakhin

Abstract

The methods of processing digital samples of complex structure signals with unknown parameters are considered. With the use of algebraic methods, the following tasks are sequentially solved: clock synchronization, determining the range of carrier frequencies, the multiplicity of phase modulation, and obtaining a stream of information bits. The methods for improving the quality of processing digital samples of signals based on solving special overdetermined systems of linear equations are proposed. The estimation of efficiency of the offered method is carried out by an imitation statistical modeling. The advantages of the proposed methods of signal processing for the telecommunications and radio monitoring systems are shown.

Keywords: orthogonal frequency-division multiplexing, digital sampling, linear algebraic equations, correlation convolutions

1. Introduction

The construction of effective information transmission systems is inextricably linked with the problem of intensifying the usage of the time and frequency-energy resource of communication channels. One of the ways to solve this problem is using the complex signals with combined types of modulation in combination with the methods of spectrum narrowing and noise-resistant coding [1–5]. In this connection, the structure of the signals used to transmit information is becoming complicated, and, consequently, the algorithms of their processing are becoming complicated as well.

The most promising type of signal code constructions in wireless networks is orthogonal frequency-division multiplexing (OFDM) [2–5]. The basic idea of building such signals is arranging a set of mutually orthogonal frequency subchannels so that, on the one hand, one subchannel does not interfere with the other and, on the other hand, the spectra of the subchannels overlap. Due to the orthogonality of the linear subchannels, each of them can be considered independently of the others. Errors caused by the interference in one of the subchannels do not lead to errors in the other. As a result, only a small part of the transmitted information is distorted. Error-correcting coding being used, the errors can be corrected. The structure of signals with multiple simultaneously operating subcarrier frequencies has well established itself in conditions of heterogeneity of the propagation medium. In recent years, the capabilities of systems with OFDM signals have evolved significantly. Such signals began to be used in a wide variety of telecommunication systems operating in different radio frequency bands.

The complex structure of such signals and the a priori uncertainty of the channel properties cause significant difficulties in solving the problems of radio control and radio monitoring [6–11]. Similar tasks also arise when processing images and signals in biotechnical systems [12–16]. A distinctive feature of such tasks is the absence of data on the structure and informative parameters of the measured signals. This information should be obtained from the results of the study, with high accuracy and as soon as possible. Therefore, the tasks of developing mathematical methods for analyzing complex signals based on digital measurement sequences are highly relevant. This is evidenced by the intensive development of special software systems [17].

2. Mathematical model of OFDM signals

For correct choice of the methods for digital analysis of the primary parameters of OFDM signals, a brief description of their basic properties is necessary. Arbitrary OFDM signal $S_j(t)$ on j-th modulation interval T_p is formed by algebraic summation of the several harmonic oscillations of the same amplitude. Each of the oscillations has m options of modulation phase shift. The value m determines the multiplicity of the used phase (PM) modulation and corresponds to the base of the numerical source code. Commonly, $m = 2^k$ where k is the number of binary symbols (bits) represented by the elementary signal on one modulation interval. When using relative phase coding and a unit value of the amplitude of the oscillation subcarriers, the mathematical model of the signal can be represented as the following sequence:

$$S_j(t) = \sum_{i=0}^{n_f-1} \sin\left\{2\pi\left(f_0 + \frac{i}{T}\right) \cdot \left(t - T_p\left\lfloor\frac{t}{T_p}\right\rfloor\right) + \varphi_{j,i}\right\}, \tag{1}$$

where t is the current time, f_0 is the lowest subcarrier frequency in the signal spectrum, $T = 1/\Delta f$ is inverse of the minimum subcarrier spacing Δf, n_f is the number of frequencies used, and $\varphi_{j,i}$ is the value of the manipulation angle of i-th fluctuation on j-th modulation interval. This angle can take one of the m values depending on the manipulation code used. The informative features in the signal described by model (1) are relative phase jumps in carrier frequencies. These jumps are measured for each of the frequency subcarriers separately: $\varphi_{j,i} - \varphi_{j-1,i}$, $i = 0, \ldots, n_f - 1$. The time parameters of the modulation interval used in model (1) are tied by the relation

$$T_p = T + \Delta T = \frac{1}{\Delta f} + \Delta T, \tag{2}$$

where ΔT is the duration of the prefix part of the signal. The prefix part (hereinafter the prefix) is a repeating (to the exact sign) initial part of the signal added at the end of the modulation interval T_p. Prefix structure is used to facilitate synchronization in the presence of channel irregularities. For OFDM synchronization violations, the signal can be correctly received and processed for any segment with the duration of T within the full modulation interval T_P. Commonly, the choice of the prefix duration on the modulation interval corresponds to the ratio $\Delta T = (0,1 \div 0,5)T$. The sign of the prefix depends on the value of the following parameter:

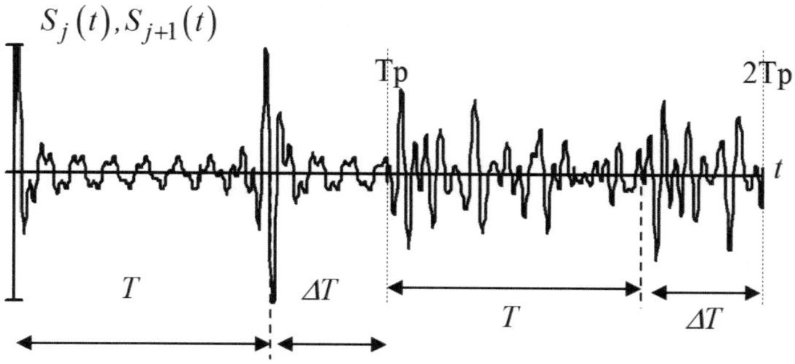

Figure 1.
Example of the OFDM signal.

$$P = f_0 \text{mod}(\Delta f). \tag{3}$$

For existing OFDM standards, parameter (3) can take two values that determine the sign of the cyclic prefix: when $P = 0$ the prefix is positive and when $P = \Delta f/2$ inverse. **Figure 1** gives a qualitative idea of the form of the signal envelope constructed in accordance with model (1) on two adjacent modulation intervals with $n_f = 16$ and $T_p = 1, 47 \cdot T$. At the end of each of the intervals T_p, the inverse cyclic continuation of the signal with duration ΔT is located, which repeats, up to a sign, the shape of the initial segment of the signal on the modulation interval. For the example in question, $P = \Delta f/2$; therefore the prefix part is the inverse of the initial part of the signal.

3. The main stages of the structural analysis of OFDM signals

A comprehensive analysis of the properties of complex signals is advisable to implement on the basis of phased processing. At each stage, only a part of the signal parameters is determined. Given the fact that OFDM signals contain a prefix, it is advisable to use the correlation method for determining structural time parameters T_p and T at the first stage. This technique is based on the principle of "sliding" time window. This makes it possible to determine the following parameters of an OFDM signal: the value of the orthogonality interval, the duration of the modulation interval, and the value of the frequency spacing between the channels.

At the second stage of the analysis, the tasks of determining the number and the values of service and information channel frequencies, as well as, the signal phase demodulation.

The two-stage processing results in the possibility to extract an information flow from signals of an a priori unknown structure without using traditional fast Fourier transform (FFT) algorithms.

3.1 Correlation method for determining the time parameters of OFDM signals

We propose a correlation method for determining structural time parameters T_p and T. The basis is the "sliding time window" principle. The most probable value of the time interval between the most correlated segments (with the "+" or "−" sign) of the segments from the digital sample of signal measurements is determined. The assessment of T_p—the most likely period of the emergence of "bursts" of

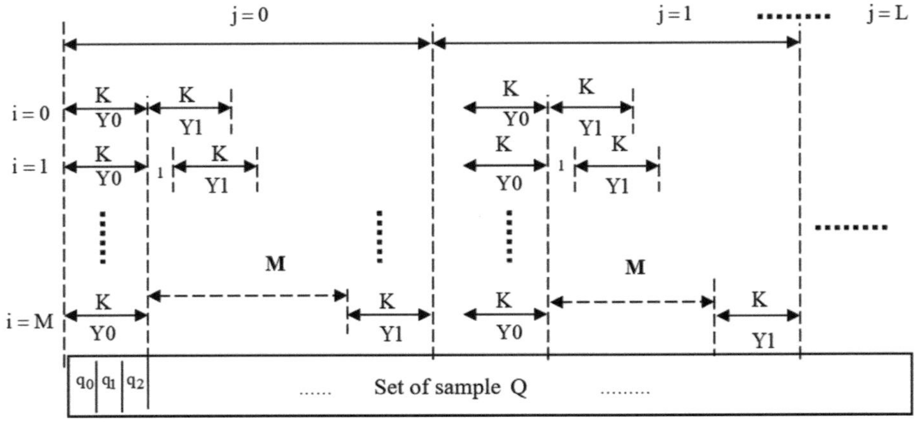

Figure 2.
Calculation scheme.

correlation in the process of moving the viewing window on the samples of the array of measurements $Q = \{q_0, q_1, \dots\}$—is determined as well. The scheme of the calculation procedure is presented in **Figure 2**. For the correlation analysis, the two vectors, each containing K elements of array Q in two non-overlapping time observation windows of the signal, are formed:

$$Y0 = \left\{q_j, q_{j+1}, \dots q_{j+K-1}\right\}, \\ Y1 = \left\{q_{j+i+K}, q_{j+i+K+1}, \dots q_{j+i+2K-1}\right\}, \tag{4}$$

removed from each other by i, $i = 0, \dots, M$. The position of the second time window corresponding to the vector $Y1$ is determined by the successive change in the offset index $i = 0, \dots, M$ that ensures its "slip" along the signal sample Q at each of the values $j = 0 \dots L$.

For a wide range of analyzed signals, for example, for the 0.3–3.4 KHz frequency band with minimum quality ADC, the most universal limits of the values of these parameters, resulting in a quick and accurate assessment, are $K = 10 \div 30$, $M = 200 \div 300$, and $L = 1000$.

At each value of index j (moving the window slip area), the $M + 1$ dimensional vector is being formed:

$$V_j = \left\{v_0^j, v_1^j, \dots, v_M^j\right\}, \tag{5}$$

the elements of which are the coefficients of mutual correlation of vectors $Y0$ and $Y1$. The calculations (according to **Figure 2**) are performed after centering and normalizing the vectors by the formulas

$$Y0_N = \left[Y0_j - \frac{1}{K}\sum_{i=0}^{K-1} Y0_i\right] \cdot \left[\frac{1}{K}\sum_{i=0}^{K-1} Y0_i\right]^{-1}; \\ Y1_N = \left[Y1_j - \frac{1}{K}\sum_{i=0}^{K-1} Y1_i\right] \cdot \left[\frac{1}{K}\sum_{i=0}^{K-1} Y1_i\right]^{-1}. \tag{6}$$

The resulting vectors $Y0_N$ and $Y1_N$ in normalized space have the same length, equal to $\sqrt{K}$, and the cosine of the angle between these vectors is equal to the cross-correlation coefficient:

$$cos(Y0_н, Y1_н) = \sum_{i=0}^{K-1} Y0_{нi} \cdot Y1_{нi} = r(Y0_н, Y1_н). \tag{7}$$

Since the prefix is a repetitive (up to sign) part of the OFDM signal, ideally the correlation coefficient between these parts is ± 1.

Figure 3 shows the distribution of the values of the elements of the vector V_j calculated according to a specific implementation OFDM signal (16 carrier frequencies, a modulation rate—75 bauds) at $M = 200$. The presence of pronounced extreme values which are close in magnitude to unity is obvious. According to the results of calculations when $j = 0...L$, the two new vectors $V1 = \{v1_0, v1_1, ..., v1_L\}$ and $V2 = \{v2_0, v2_1, ..., v2_L\}$ are formed. Their elements are calculated according to the rules

$$\begin{matrix} v1_j = match\left[min\left(V_j\right), V_j\right]_0 + K; \\ v2_j = match\left[max\left(V_j\right), V_j\right]_0 + K; \end{matrix} \quad j = 0...L. \tag{8}$$

Here the function $match[x, X]$ calculates the indices of the elements of the vector X equal to x, where the index 0 in the function (8) indicates a selection of the element with a minimum sequence number, if there are several such elements in the vector. The elements of the vector $V1$ represent the number of sampling intervals that fit between the initial elements $Y0$ and $Y1$ with minimal (negative) correlation on j-th step of moving the observation window. Accordingly, the elements of the vector $V2$ are calculated for the maximum (positive) correlation of the vectors $Y0$ and $Y1$. Simultaneous determination of the maximum and minimum is necessary to reveal the value of function (3). It is obvious that the elements of the vectors $V1$ and $V2$ defined by expression (8) can take values only in the range $K...K + M$. To study the statistical distribution of the values of the elements, the histograms for the elements of the vectors $V1$ and $V2$ are formed:

$$\begin{matrix} H1 = \{h1_0, h1_1, ..., h1_{K+M}\}, \\ H2 = \{h2_0, h2_1, ..., h2_{K+M}\}. \end{matrix} \tag{9}$$

Obtaining distributions (9) gives an opportunity to estimate the most likely value Num—the number of sampling intervals between the initial measurements of the segments of the digital sample Q with maximal (positive or negative) correlation:

$$Num = \begin{cases} \underset{\langle max(H1) \geq max(H2) \rangle}{match[max(H1), H1]_0;} \\ \underset{\langle max(H1) < max(H2) \rangle}{match[max(H2), H2]_0.} \end{cases} \tag{10}$$

The value Num determines the number of sampling intervals that fit on the orthogonality interval of the signal T. It gives an opportunity to find two

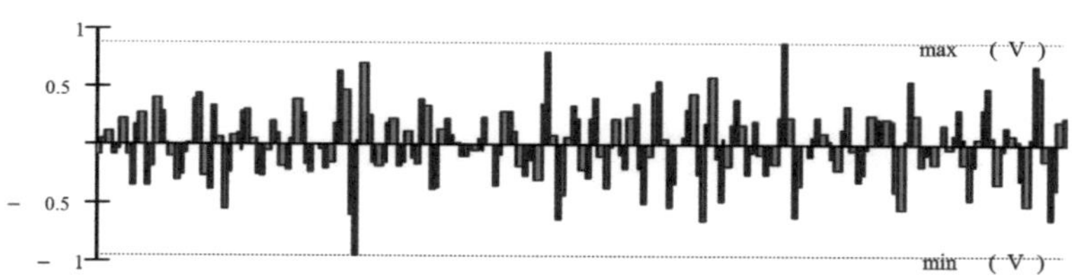

Figure 3.
Cross-correlation coefficients.

interrelated OFDM signal parameters: the orthogonality interval and the minimum carrier frequency spacing:

$$T = \frac{Num}{f_{\partial}} \text{ and } \Delta f = T^{-1}. \tag{11}$$

Besides, choosing the corresponding method of detecting the *Num* value according to the condition specified in function (10) automatically determines the value of function (3) and, consequently, the ratio between the frequency parameters f_0 and Δf. If the value *Num* is determined by the first line of expression (10), then $f_0 = \left(k + \frac{1}{2}\right) \cdot \Delta f$. Otherwise the minimal carrier frequency is multiplied to the spacing of the carrier frequencies: $f_0 = k \cdot \Delta f$ where k is any positive integer.

As a result of processing histograms (9) according to function (10) based on the values of function (3), only one of the vectors $V1$ or $V2$ is left for the further analysis, hereinafter denoted V^*. This is possible because the prefix repetition sign is defined. Based on the elements of the vector V^*, another vector $V3$ is formed for the analysis:

$$V3 = match[Num, V^*]. \tag{12}$$

The elements of this vector are equal to the numbers of the elements of the vector V^* in which the numbers *Num* are located. The feature of the vector V^*, provided that the analyzed signal belongs to the OFDM class, is that it contains a sequence of periodic series of numbers which are close or equal to *Num*. Therefore, the values of the elements of $V3$ in order of increasing their indices will be the segments (series) of an ordinary positive integer sequence with some gaps in the sequence. Small gaps can be observed inside the series too. A possible approximation of the sequence of the elements of $V3$ is illustrated by the following expression:

$$V3 = \left\{ \underbrace{11, 12, 14, 15, 16}_{\text{series№1}}, \underbrace{105, 107, 108, 110}_{\text{series№2}}, \ldots, \underbrace{620, 621, 622, 623, 624}_{\text{series№}n_c} \right\}$$

The length of series of consecutive numbers (position numbers) may differ due to measurement errors, features of the signal envelope, and rounding during calculations. However, when the vector length is sufficient, averaging results converges to the true estimation in accordance with the law of large numbers. To exclude "fragmentation" of the series, small gaps between adjacent numbers of the series must be ignored. It has been empirically found that in most cases, the number N_i should be considered to belong to the current series of numbers if $N_i - N_{i-1} \le \varepsilon$ where $\varepsilon = 3 \div 5$. In general, the structure of the vector $V3$ can be depicted as shown in **Figure 4**.

In **Figure 4**, $v3_{n_j}$, $v3_{k_j}$ are the initial and final elements of the j-th series of the consecutive numbers in the vector $V3$, and n_c is the total number of the identified

$$\left\{ \underbrace{\boxed{v3_0, \ldots, v3_{k_1}}}_{\text{series №1}}, \underbrace{\boxed{v3_{n_2}, \ldots, v3_{k_2}}}_{\text{series №2}}, \ldots, \underbrace{\boxed{v3_{n(n_c-1)}, \ldots, v3_{k(n_c-1)}}}_{\text{series №}(n_c-1)}, \underbrace{\boxed{v3_{n(n_c)}, \ldots, v3_{k(n_c)}}}_{\text{series №}n_c} \right\}$$

Figure 4.
Example of the V3 series structure.

series. Using the presented structure and values of the elements of the vector $V3$, it is possible to determine the number of sampling intervals that fit between adjacent pairs of mutually correlated segments of signal measurements, i.e., a period of "bursts" of correlation:

$$Num1 = \frac{\left(v3_{k_{(n_c-1)}} - v3_{k_2}\right) - \left(v3_{n_{(n_c-1)}} - v3_{n_2}\right)}{2(n_c - 2)} \tag{13}$$

Using the obtained value $Num1$, we can determine the average value of the modulation interval T_p and, therefore, the average modulation rate W:

$$T_p = \frac{Num1}{f_\partial}, W = T_p^{-1} \tag{14}$$

For the final determination of the time-frequency structure of the signal, we must find the number of carrier frequencies n_f and the vector of their nominal values $F = \left\{f_0, \ldots, f_{n_f-1}\right\}$. This can be done on the basis of previously obtained values T_p, T, Δf when the position of the element of the array Q corresponding to the beginning of the first full modulation interval is determined correctly. The beginning of a reliably identified clock interval could be most correctly associated with the beginning of the second series of maximal responses of correlators in the vector $V3$, since, due to the randomness of the beginning of the observation, the first series may be incomplete. It should be taken into account that the beginning of a series of maximal responses of correlation of the segments from K samples must appear before the next modulation interval actually begins. Therefore, to fall within the interval with the duration T_p (taking into account that demodulation can be performed on any segment T within T_p), it is necessary to add the number $K/2$ to the starting sample, at least. Then the beginning of the modulation interval can be assumed to coincide with the next element number in the sequence

$$n_T^0 = round\left\{\left[v3_{n_2} mod\left(\frac{f_\partial}{W}\right)\right] + \frac{K}{2}\right\} \tag{15}$$

Here $round(x)$ is the rounding function to the nearest integer. The lowest possible frequency f_0 in the group of carrier frequencies is determined by the value of function (3) and the fulfillment of the corresponding condition in function (10):

$$f_0 = \begin{cases} \frac{1}{2}\Delta f, & npu \quad max(H1) \geq max(H2); \\ \Delta f, & npu \quad max(H1) < max(H2). \end{cases} \tag{16}$$

The maximal number of subcarrier frequencies (or half the number of quadrature components) that can fit in the channel band F_{ef} is

$$n_{fmax} = round\left(\frac{F_{ef} - f_0}{\Delta f} + 1\right) \tag{17}$$

3.2 Determining the amount and nominal values of subcarrier frequencies of OFDM signals

The correlation method, considered above, allows making a reliable assessment of the main structural parameters of OFDM—T and T_P. The value of $\Delta f = T^{-1}$ uniquely defines the spacing of adjacent subcarrier frequencies. The minimal value of the subcarrier frequency and the maximal possible number of subcarriers placed within the signal bandwidth n_{fmax} are determined from Eqs. (16) and (17).

The number of samples N, taken into account when analyzing a signal on one modulation interval, as well as the harmonic quadrature $(2 \cdot n_{fmax})$, define the dimensions of the matrix of the linear algebraic equations system (SLAE) which can be compiled and solved to estimate the frequency range. Depending on the ratio of the vertical and horizontal dimensions of the matrix of coefficients, the system of equations can be overdetermined $(N > 2 \cdot n_{fmax})$, determined $(N = 2 \cdot n_{fmax})$, or underdetermined $(N < 2 \cdot n_{fmax})$. The simplest one is the $(N = 2 \cdot n_{fmax})$ case because then the SLAE is a joint one almost every time. The number of equations that matches the number of used elements of the digital sample Q equals to the number of unknowns $(2 \cdot n_{fmax})$ determining the amplitudes of quadrature components in the spectrum of carrier frequencies OFDM. For the correct solution of SLAU $(2 \cdot n_{fmax})$, uniformly spaced sample counts Q starting from the point of beginning of the observation of the first complete clock interval of signal n_T^0 should be selected on i-m modulation interval. For this the following rule is used:

$$n_T^i = n_T^0 + round(i \cdot \Delta) \text{ where } \Delta = T_P/t_\partial. \tag{18}$$

Square matrix of coefficients for unknown SLAE with size $(2 \cdot n_{fmax}) \times (2 \cdot n_{fmax})$ composed for quadrature components of subcarrier frequencies is formed according to the rule

$$\begin{aligned} A_1 &= \|a_{i,j}\|, \quad i,j = 0, \ \ldots, \ (2 \cdot n_{fmax} - 1); \\ a_{i,j} &= \text{Sin}\left[\omega_j \cdot t_i\right], \qquad 0 \le j \le n_{fmax} - 1; \\ a_{i,j} &= \text{Cos}\left[\omega_j \cdot t_i\right], \qquad n_f \le j \le 2 \cdot n_{fmax} - 1; \end{aligned} \tag{19}$$

where $\omega_j = \omega_{(j+n_{fmax})} = 2\pi(f_0 + j \cdot \Delta f), \quad j = 0, \ \ldots, \ n_{fmax}, t_i = n_T^i + i \cdot t_\partial$.

The column matrix of free members is formed as a vector of signal measurements on the duration of one orthogonality interval:

$$B_1 = \left\{b_0, \ldots, b_{(2 \cdot n_{fmax} - 1)}\right\}; b_i = q_i, \quad i \cdot t_\partial \in T_P{}^i \tag{20}$$

Normal solution of normally defined SLAE

$$A_1 \cdot X_1 = B_1 \quad \Rightarrow \qquad X_1 = A_1^{-1} \cdot B_1 \tag{21}$$

gives an estimation of the amplitude vector of quadrature components $X_1 = \left\{x_0^1, \ldots, x_{(2 \cdot n_{fmax} - 1)}^1\right\}$ which corresponds to the permissible values of carrier frequencies.

On the basis of this solution, it is possible to determine the power distribution vector of the signal between the harmonic oscillations of the carrier frequencies:

$$Y = \left\{y_0, \dots, y_{\left(n_{fmax}-1\right)}\right\}, \quad y_i = \left(x_i^1\right)^2 + \left(x^1_{\left(i+n_{fmax}\right)}\right)^2, \quad i = 0, \quad \dots, \quad \left(n_{fmax} - 1\right) \tag{22}$$

The case of insufficiently defined SLAE $\left(N < 2 \cdot n_{fmax}\right)$ is interesting for analyzing small samples of the signal. To solve such a SLAE, the pseudoinverse matrix method Moore-Penrose can be used. It is known that there is a normal solution of an underdetermined SLAE, and it is the only one. It is found by $X_1 = A^+{}_1 \cdot B_1$ where $A^+{}_1$ is the Moore-Penrose pseudoinverse matrix of size $2 \cdot n_{fmax} \times 2 \cdot n_{fmax}$ which is determined by the ratio $A_1 \cdot A^+{}_1 \cdot A_1 = A_1$. In practice $A^+{}_1$ can be found by the formula

$$A_1^+ = C^+ \cdot D^+ = C^* \cdot (C \cdot C^*)^{-1} \cdot (D \cdot D^*)^{-1} \cdot D^*.$$

The representation of the matrix A_1^+ in the form of a product of two matrices with the size of $N \times r$ and $r \times N$ is used:

$$A_1 = D \cdot C = \begin{pmatrix} d_{1,1} & \dots & d_{1,r} \\ \vdots & \ddots & \vdots \\ d_{N,1} & \cdots & a_{N,r} \end{pmatrix} \cdot \begin{pmatrix} c_{1,1} & \dots & c_{1,N} \\ \vdots & \ddots & \vdots \\ c_{r,1} & \cdots & c_{r,N} \end{pmatrix}.$$

With various skeletal decompositions of the matrix A, the same solution for A^+ which can be written in the form $X_1 = A^+{}_1 \cdot B_1$ is derived. It is a pseudosolution giving a zero residual: $\|X_1 - A^+{}_1 \cdot B_1\| = 0$.

The case of $\left(N > 2 \cdot n_{fmax}\right)$ is the most advantageous for the maximal recording of signal information on the modulation interval. Due to using additional signal measurements from the sample Q, the system which contains more equations with the same number of unknowns is formed. To form the matrix A_2 and the vector B_2, the maximal number of signal measurements determined by $Num \approx T_P / t_\partial$ on the duration T_P is used:

$$\begin{aligned} A_2 &= \left\|a_{i,j}\right\|, \quad i = 0, \quad \dots, \quad (Num - 1), \quad j = 0 \dots \left(2 \cdot n_{fmax} - 1\right); \\ a_{i,j} &= \text{Sin}\left[\omega_j \cdot t_i\right], \qquad 0 \le j \le n_{fmax} - 1; \\ a_{i,j} &= \text{Cos}\left[\omega_j \cdot t_i\right], \qquad n_{\max} \le j \le 2 \cdot n_{fmax} - 1; \end{aligned} \tag{23}$$

$$B_2 = \left\{b_0, \quad \dots, \quad b_{(Num-1)}\right\}, \quad b_k = q_k, \quad k = 0, \quad \dots, \quad (Num - 1). \tag{24}$$

SLAE has the form

$$A_2 \cdot X_2 = B_2 \tag{25}$$

and, as a rule, has many solutions. To select the only one, we need to use some criteria. In practice, the maximum likelihood criterion is used more often. In the case of a normal distribution of the vector B_2, it is equivalent of the least square's criterion:

$$X_2^* = \left(A_2{}^T \cdot A_2\right)^{-1} A_2{}^T \cdot B_2 \tag{26}$$

An approximate solution of system (26) gives a more accurate result than a strict solution of system (25). The noise immunity of the solution is achieved by averaging the disturbing effect of interference when the number of signal measurements exceeds the required minimum. The obtained vector of amplitudes of the quadrature components X_2^* as well as X_1^* gives a possibility to calculate the power distribution signal in carrier frequencies using expression (22), wherein x_i^2 is used instead of x_i^1.

For any type of SLAE, determining the actual list of carrier frequencies in the OFDM spectrum is performed by comparing the elements of power distribution histograms with a threshold value. The obtained nominal values of frequencies determine the last structural time-frequency parameter of the analyzed signal—the vector of working subcarrier frequencies F.

Thus, the previously obtained signal parameters T_p, T, Δf, W, and the obtained in this subsection vector F identify completely their structural properties and make signal demodulation possible.

4. Group algebraic demodulation

4.1 Determination of phase modulation multiplicity

OFDM standards as an informative parameter on subcarriers assume the use of relative phase shift keying. The number of variants of phase angles can vary from 2 to 16 or more [2–5]. As a rule, absolute phase modulation methods are not applied because they are critical to the correctness of the subcarrier phase restoration, which is highly dependent on the accuracy of determining the beginning of the modulation interval. In case of digital demodulation, when a sample of signal measurements is formed for time points asynchronously with respect to a periodic sequence of modulation interval boundaries, the usage of absolute phase modulation methods is practically impossible, since it is theoretically impossible to provide absolutely accurate clock synchronization of modems. Therefore, as an axiomatic assumption for developing the stages of secondary analysis, the hypothesis of using the relative phase coding methods is adopted. Depending on the multiplicity of the applied modulation method $m = 2^{\alpha}, \alpha = 1,\ 2,\ 3, \cdots$ is an integer, during the transition from interval to interval; on each of the carriers, a phase jump occurs, the value of which determines the corresponding combination of information symbols. Jumps occur even in the case of complete coincidence of combinations of symbols on adjacent intervals.

The general idea of a preliminary analysis of the PM multiplicity is to statistically identify the number of observed fixed values of the phase of harmonic oscillations in carrier frequencies. To perform the analysis, let us use the algebraic method for the case of overdetermined SLAEs considered in the previous section. Let *Num* be the maximum number of measurements performed on the modulation interval and n_f the number of carriers detected during the primary analysis. Then the degree of overdetermination of SLAE is calculated from the expression

$$K_{\Pi} = \frac{Num}{2 \cdot n_f} > 1 \tag{27}$$

In accordance with the findings, the increase of K_{Π} leads to the increase in the noise immunity of the system solution. SLAE, compiled for the entire spectrum of subcarriers, is determined by a rectangular (size $Num \times 2 \cdot n_f$) matrix of coefficients for the following unknowns:

$$A = \left\| a_{i,j} \right\|;\ a_{i,j} = \begin{cases} Sin\left(2\pi f_j (i \cdot \Delta T)\right), & i = 0, \dots, Num - 1, \quad j = 0, \dots, n_f - 1; \\ Cos\left(2\pi f_j (i \cdot \Delta T)\right), & i = 0, \dots, Num - 1, \quad j = n_f, \dots, 2 \cdot n_f - 1 \end{cases}, \tag{28}$$

as well as a column matrix of *Num* signal measurements

$$B = \{b_0, \ldots, b_{Num-1}\}, \quad b_i = y_{J+i}, \quad J = n_T^0 + round\left(N_u \frac{f_\partial}{W}\right) \tag{29}$$

Here $f_\partial = (t_\partial)^{-1}$ is the sampling frequency of the digital signal sample (t_∂ is the sampling interval), $Y = \{y_0, \ldots, y_k\}$ is the digital sample of measurements of the OFDM signal, $F = \left\{f_0, \ldots, f_{n_f-1}\right\}$ is the vector of nominal values of the detected subcarrier frequencies, N_u is the number of the analyzed modulation interval, W is the modulation rate in bauds, and n_T^0 is the initial measurement number (in the digital signal sampling), corresponding to the reliably detected beginning of the first complete modulation interval. The value n_T^0 is one of the results of the stage of the primary correlation analysis using the "sliding time window" method discussed in the previous section.

To find the solution of the overdetermined SLAE (28), (29) the method of least squares is used to solve the following system:

$$A^T \cdot A \cdot Z = A^T \cdot B \tag{30}$$

where A^T is the transposed matrix (28) and $Z = \left\{z_0, \ldots, z_{n_f-1}, z_{n_f}, \ldots, z_{2n_f-1}\right\}$ is the vector of the amplitudes of the harmonic quadrature components for all n_f subcarriers of the signal frequencies.

The decision vector Zcontains the amplitudes of sines (first n_f elements) and cosines (last n_f elements) in frequencies determined by the vector F. It gives an opportunity to calculate the elements of the phase vector (expressed in degrees) of harmonic oscillations $\Phi = \left\{\phi_0, \ldots, \phi_{n_f-1}\right\}$ in the subcarrier frequencies:

$$\phi_i = mod\left[\left(\frac{180}{\pi} Atan\left(z_i, z_{i+n_f}\right) + 360\right), 360\right], \quad i = 0, \ldots, n_f - 1 \tag{31}$$

In this expression, the following functions are used:

- $mod(a, b)$—to calculate the value of the number a by modulo b
- $Atan(x, y)$—to calculate the angle in radians from the interval $[-\pi, \pi]$, between the axis $0x$ and the vector drawn from point (0,0) to point (x, y)

The process of calculations by expressions (30) and (31) is the absolute phase demodulation of discrete measurements on an interval T located within the segment of modulation interval T_P. It should be noted that with relative phase shift keying, the calculated absolute values of the phases do not matter for correct demodulation of the bit stream. Therefore, in this case, the error in determining the beginning of the orthogonality interval T within the modulation interval T_P is not significant. The calculation of the sampling number of the next interval is based on the known modulation rate W, the starting number of the initial interval n_T^0, and the sampling frequency F_∂ and performed by using the rounding formula from expression (29).

The factor that reduces the effectiveness of group digital demodulation is the asynchrony of the sampling procedure when a sample of signal measurements is being obtained with the duration of repeated demodulation intervals. In general, the value $T/t_\partial = f_\partial \cdot T$ is not an integer. This leads to a variation of the value of "indentation" dT_i from the beginning of the i-th modulation interval. This phenomenon is illustrated in **Figure 5**. Differences in the values of dT_i lead to the fact that the phase error periodically repeated on a combination of several adjacent modulation intervals is added to the correct solution of the system. The significant "blurring" of phase increments in subcarrier frequencies, as well as the departure of their average values from standardized nominal values, is an external manifestation of this asynchronous effect. Therefore, the phase errors of asynchronous sampling are an additional source of disturbing noise, which must be taken into account and, if possible, compensated.

A known correlation between magnitudes W and t_∂ allows us to estimate the magnitude of these errors. On an arbitrary i-th interval, the error in determining the initial measurement is

$$dT_i = t_\partial \left[i\frac{f_\partial}{W} - round\left(i\frac{f_\partial}{W} \right) \right] \quad (32)$$

Obviously, for different subcarrier frequencies, the same time shift dT will cause phase determination errors which are different in magnitude. By knowing the range of carrier frequencies of the vector F, we can precisely calculate the corrections to the calculations for the current phase of the L-th carrier frequency on i-th modulation interval as follows:

$$d\phi_{L,i} = 2\pi \cdot f_L \cdot dT_i = 360 \cdot f_L \cdot dT_i \quad (33)$$

The multiplicity of the phase modulation used in the observed signal is uniquely determined by the minimal difference between the closest values of the vector Φ elements calculated on the basis of expression (31), taking into account corrections (32) and (33). Under the conditions of a priori uncertainty, the multiplicity of relative PM (RPM) applied in the subcarrier frequencies can be detected only by a statistical method based on an analysis of a series of SLAE solutions from the sample of consecutive modulation intervals which has a sufficiently large size N. For this the vector is created:

$$\Psi = \left\{ \psi_0^0, \dots, \psi_{N-1}^0;\ \psi_0^1, \dots, \psi_{N-1}^1; \dots\dots; \psi_0^{n_f-1}, \dots, \psi_{N-1}^{n_f-1} \right\} \quad (34)$$

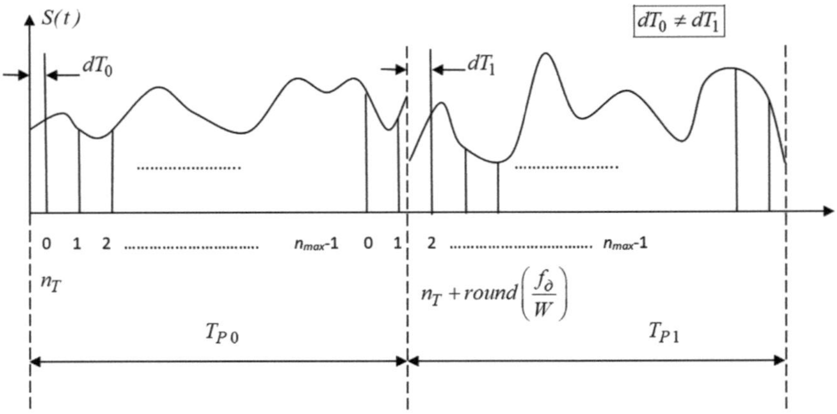

Figure 5.
Asynchronous sampling illustrations.

The vector contains N series of n_f elements. Each element in the ψ_i^j series is the phase increment measurement of the$j$-th subcarrier oscillation on the i-th modulation interval and is calculated (in degrees) as follows:

$$\psi_i^j = mod\left[\left(\phi_{i+1}^j - \phi_i^j\right), 360\right] \tag{35}$$

where ϕ_i^j is the absolute phase measurement of the j-th subcarrier on the i-th modulation interval. Since the multiplicity of RPM in all frequencies must be the same, the vector Ψ can be considered as a single statistic characterizing the entire OFDM signal. Based on the vector of the statistics of phase increments Ψ, we form a vector histogram $\overline{\Psi}$, the elements of which are proportional to the frequency of occurrence of various increments. When rounding calculations up to 1°, the histogram $\overline{\Psi}$ will contain 360 elements:

$$\overline{\Psi} = \{\overline{\psi}_0, \dots, \overline{\psi}_{359}\}, \quad \overline{\psi}_i = n\psi_i, \tag{36}$$

where $n\psi_i$ is the number of elements of the vector Φ that fall inside the interval $[i - 0,5;\ i + 0,5]$.

A visual analysis of the histogram (36) lets us reliably identify the type of modulation used. However, the automatic analysis requires the formal mathematical one. The cross-correlation which is the most reliable mathematical tool for calculating the degree of similarity of two functions should be the natural basis of such algorithm. A tuple of reference functions for calculating mutual correlation with a histogram vector $\overline{\Psi}$ can be obtained using Gaussian probability density function based on the following expression:

$$GF(m, w, d) = \frac{1}{\sigma\sqrt{2\pi}} \sum_{i=0}^{m-1} exp\left(-\frac{\left(w - d\frac{360}{m}\right)^2}{2\sigma^2}\right), \tag{37}$$

where m is the multiplicity of RPM, w the phase value in degrees, and d the offset parameter of the "reference comb" of the histogram maxima on the interval $[0 - 360^{\circ}]$. This expression has no probabilistic sense, since the Gaussian distribution is used only as the convenient analytical definition of the multimodal structure of reference functions. An example of generation of references for the multiplicity of modulation $m = 2, 4, 8$ *u* 16 is presented in **Figure 6**.

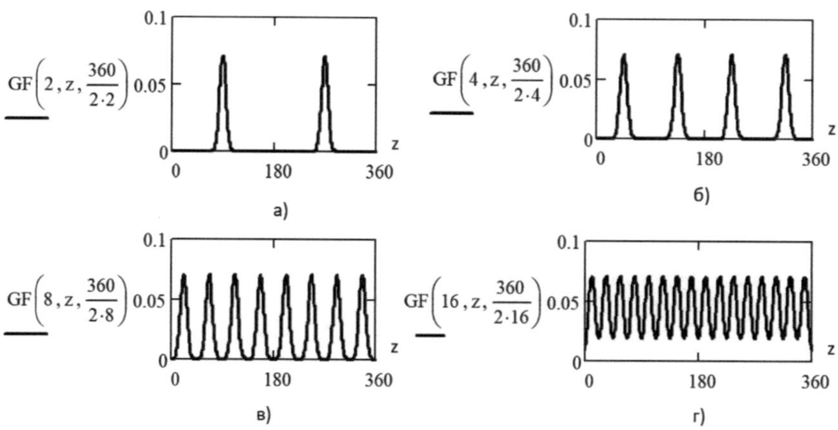

Figure 6.
Tuple of reference functions.

The structure of the references specified by Eq. (37) ensures the symmetry of the structure of functions relative to the average value of the phase π. For the given multiplicity m based on Eq. (37), the vector x for different values of the shift of the reference comb is formed:

$$x = \{x_0, \ldots, x_{359}\}, \qquad x_i = GF(m, i, j), \quad i = 0, \ldots, 359; \quad j = 0, \ldots, 180 \qquad (38)$$

For each shift value j, the cross-correlation coefficient is calculated: $X_j = \frac{x \times \overline{\Psi}}{|x| \cdot |\overline{\Psi}|}$ where "$\times$" denotes a scalar product.

The maximal obtained value $Y_m = max(X_j)$ is taken as an indicator of the similarity of the histogram $\overline{\Psi}$ with the reference of the given multiplicity. Then the next value m is taken and calculations are repeated. After a complete exhaustion of all possible values of multiplicity, the largest calculated value Y_m is determined. The index of this element corresponds to the detected multiplicity of phase modulation m. The described computational process for calculating the correlation proximity is conveniently implemented using the mechanism of nested cycles. It is easy to program, and the numerous tests show that it guarantees the correct finding of the modulation multiplicity even under conditions of significant distortion of signal samples.

4.2 Algebraic demodulation

For the final demodulation, the SLAE of reduced dimensionality is compiled with a rectangular matrix of coefficients with unknowns:

$$\begin{array}{l} A = \left\| a_{i,j} \right\|, \quad i = 0, \ldots, Num - 1, \quad j \in N; \\ B = \{b_0, \ldots, b_{Num-1}\}, \quad b_i = y_{J+i}, \quad J = n_T^0 + round\left((N \cdot f_\partial)/W\right). \end{array} \qquad (39)$$

The solution of the system $A^T \cdot A \cdot Z = A^T \cdot B$ for several modulation intervals gives a vector of absolute values of the phases which gives a sequence of corrected phase increments after transformations. To identify the combinations of binary symbols in accordance with the obtained modulation multiplicity and the corresponding keying code, it is necessary to calculate the decision boundaries. Since all combinations of binary symbols are assumed to be equally probable and phase errors are distributed normally and symmetrically around the standard positions, then, in accordance with the maximum likelihood rule, it is advisable to use equidistant decision-making boundaries:

$$G_i = i\frac{360}{m}, \quad i = 0, \ldots m \qquad (40)$$

Then the rule for forming a bit stream at the output of the investigated example of a digital demodulator when using, for example, four phase angles on i-th modulation interval for one carrier frequency can be written as a set of conditions for forming a pair of binary symbols:

$$s_{2 \cdot i}, s_{2 \cdot i+1} = \begin{cases} 1, 0; & if \quad G_0 \le \Psi_i < G_1; \\ 0, 0; & if \quad G_1 \le \Psi_i < G_2; \\ 0, 1; & if \quad G_2 \le \Psi_i < G_3; \\ 1, 1; & if \quad G_3 \le \Psi_i < G_4; \end{cases} \qquad i = 0, 1, 2, \ldots \qquad (41)$$

	Parameter	Value		
		Samples	**Seconds**	**Hertz**
Stage 1	Period of orthogonality	73	0.009	—
	Subcarriers distance	—	—	110
	Duration of the OFDM character	107.083	0.013	—
	Sign of the prefixes correlation—negative			
	Starting number of the first full OFDM character	107	—	—
	Time of analysis	—	0.5	—
	Sample size of measurements	4001	—	—
Stage 2	The possible frequency grid $i \in 0...31$	—	—	$55 + i \cdot 110$
	Numbers of carriers—17			
	Nominals of carriers $j \in 0...16$			$825 + j \cdot 110$
	Number of phase angles PSK—4			
	Number and nominal value of carrier of synchronization channel =0	—	—	825
	Numbers and nominal values of information carriers $k \in 1...16$	—	—	$825 + k \cdot 110$

Table 1.
Analysis experiment results.

The full stream of transmitted characters is formed with this rule. The result of the work of the considered method of automatic technical radio monitoring of the source of OFDM signals is represented as received by binary sequences, for example,

$X_0 =$

	0	1	2	3	4	5	6	7	8	9	10	11
0	1	0	0	0	1	0	0	1	1	0	0	...

$X_1 =$

	0	1	2	3	4	5	6	7	8	9	10	11
0	1	1	1	1	0	0	1	0	1	1	0	...

and so on.

The correctness of the considered method for analyzing and demodulating the OFDM signals under conditions of partial uncertainty has been tested using both real and simulated digital measurements of signals.

Based on the considered algebraic algorithms for automatic analysis, recognition, and demodulation of signals, a software package was developed. **Table 1** presents an example of the main results of the use of the software package for the analysis and demodulation of a complex signal of the standard MIL188-110A. The moment of the beginning of a signal observation was accidental. The test signal was sampled in the low-frequency band 0.3÷3.4 kHz after spectrum transfer by decimation of measurements. In this example, the signal contained 16 informational subcarrier frequencies and 1 synchronization subchannel at a modulation rate of 75 OFDM characters/s and relative PSK-4. The sampling rate was 8000 sample/s.

From the data in the table, it follows that it took 0.5 seconds of observation to identify all the parameters of the signal and start demodulation. Note that signal processing can be done in real time or based on the results of previously stored digital samples.

5. Conclusions

The considered statistical method for analyzing the structure and demodulation of OFDM signals under conditions of a priori uncertainty of solving radio monitoring tasks has been practically tested. It has demonstrated the high accuracy of parameter identification. The relatively low computational complexity of correlation and algebraic analysis makes it possible to identify the structure and the parameters of signals practically in seconds.

The noise immunity of the analysis is achieved by solving a SLAE with rectangular overdetermined matrixes of coefficients. To eliminate phase errors due to the asynchrony of the sample relative to the clock modulation intervals, a method for calculating phase corrections is proposed. The method uses the known parameters of the time-frequency structure of the signal. The application of the phase correction method provides ideal conditions for identifying the modulation type of subcarrier oscillations. Mathematical formalization of solving the problem of determining the modulation multiplicity, based on generating the multimodal reference functions and sequential calculating the degree of mutual correlation, allows us to completely automate the process of identifying the secondary parameters which are necessary for demodulating the signals of subcarrier frequencies. The further research can be focused on the generalization of the method for any structures of mono and poly frequency signals including those with a linear frequency modulation.

Author details

Sergey G. Rassomakhin
Information Systems and Technologies Security Department, V.N. Karazin Kharkiv National University, Kharkiv, Ukraine

*Address all correspondence to: rassomakhin@karazin.ua

References

[1] Golomb SW, Gong G. Signal Design for Good Correlation for Wireless Communication, Cryptography, and Radar. Ontario. ISBN: 9780521821049: University of Southern California, University of Waterloo; 2005

[2] Shulze H, Luders C. Theory and Applications of OFDM and CDMA. Chichester: John Wiley & Sons, Ltd.; 2005. ISBN: 13 978-0-470-85069-5

[3] Mignone V, Morello A. OFDM: A novel demodulation scheme for fixed and personal receivers. IEEE Transactions on Communications, 1996. - №44. - R. 1144-1150

[4] Li Y, Cimini LJ Jr, Sollenberger NR. Robust Channel estimation for OFDM systems with rapid dispersive fading channels. IEEE Transactions on Communications. 1998;**46**(7):902-915

[5] Zhao Y, Huang A. A novel channel estimation method for OFDM mobile communication systems based on pilot signals end transform—Domain processing. Proceedings VTC. 1997;**33**: 2089-2093

[6] Stepanov A, Matveev S. Methods of Computer Signal Processing. Moscow: SOLON-Press; 2003. p. 207

[7] Ka Mun H. Automatic recognition and demodulation of digitally modulated communications signals using wavelet-domain signatures [Internet]. A Dissertation for the degree of Doctor of Philosophy. New Brunswick, New Jersey; 2010.p. 201. Available from: https://rucore.libraries.rutgers.edu/rutgers-lib/26533/PDF/1/play/ [Accessed: 28 February 2019]

[8] Dobre OA, Abdi A, Bar-Ness Y, Su W. A survey of automatic modulation classification techniques: Classical approaches and new trends. IET Communications. Apr. 2007;**1**(2):137-157

[9] Richterova M. Signal modulation recognizer based on method of artificial neural networks. In: Proc. 2005 Progress in Electromagnetics Research Symp. 2005. pp. 575-578

[10] Hong L, Ho KC. Classification of BPSK and QPSK signals with unknown signal level using the Bayes technique. Proceedings of the 2003 IEEE International Symposium on Circuits and Systems. 2003;**4**:1-4

[11] Li J, He C, Chen J. Automatic digital modulation identification basing on decision method and cumulants. In: Proceedings of 2005 IEEE International Workshop on VLSI Design and Video Technology. 2005. pp. 264-267

[12] Masciotti JM, Lasker JM, Hielscher AH. Digital lock-in detection for discriminating multiple modulation frequencies with high accuracy and computational efficiency. IEEE Transactions on Instrumentation and Measurement. 2008;**57**(1)

[13] Schnyder F, Haller C. Implementation of FM Demodulator Algorithms on a High Performance Digital Signal Processor [Internet]. Available from: https://www.veron.nl/wp-content/uploads/2014/01/FmDemodulator.pdf [Accessed: 28 February 2019]

[14] Hamidi SA, Jafari R, Moosavi Nia A, Soleimani M. Design and implementation of a DSP-based digital phase sensitive demodulation for an EIT system. In: Proceedings of the International Conference on Electrical Bioimpedance. Journal of Physics: Conference Series 224; 2010. 012147. pp. 1-4. DOI: 10.1088/1742-6596/224/1/012147

[15] Trong T. Dong Sun Direct Demodulation of Optical BPSK/QPSK Signal without Digital Signal Processing

[Internet]. Available from: https://www.radioeng.cz/fulltexts/2018/18_04_0942_0947.pdf [Accessed: 28 February 2019]

[16] Tikhonov A, Arsenin V. Methods for solving ill-posed problems. Moscow: Science. 1979. p. 287

[17] Go2DECODE. The Standalone Software for Signal Recognition, Demodulation, Decoding, Speech Detection, Signal Recording and Technical Signal Analysis [Internet]. Available from: https://www.hik-consulting.pl/files/go2DECODE-Brochure-V1.0-2012-05.pdf [Accessed: 28 February 2019]

Chapter 3

Integrated Electro-Optics Modulator

Yufeng Tao

Abstract

Electro-optic modulation (EOM) is an essentially important optical manipulation for on-chip photonics, optical communication and optical sensing. With emerging demands on efficient, broadband electro-optic modulation, the high-performance, integrated electro-optic modulation becomes indispensable. By manipulating phase or amplitude of optical field, optical information will be coded/modulated for communication or modulation. Through advanced micro/nano fabrication technique, the electro-optics crystal could be cut into the required volume/shape as specific, integrated modulator, waveguide or meta-surface for nano-photonic applications, paving a solid way for the imminent nano-photonic devices. Herein, the basic electro-optics effects, opto-electronic applications, methods of fabrication/integration, and future prospect of lithium niobate crystal are discussed or introduced. Demonstrations of box-sealed EOM, in-fiber EOM and the fabricated lithium niobate waveguides on substrate will be found here.

Keywords: electro-optics modulator, lithium niobate, micro/nano fabrication, nanometric interferometric sensing, on-chip photonics

1. Introduction

The emerging microwave photonics, optical communication network, photonic quantum computation and optical sensing require advanced electro-optics modulators with wide electrical modulation bandwidth (over 100 GHz), ultra-low optical insertion loss (<3 dB), scalable size, low half-wave drive voltage, ultra-fine signal quality, massively-producible fabrication ability and easy integration on various of insulator platforms [1–3]. Therefore, electro-optic modulators (EOMs) based on nonlinear optical materials such as lithium niobate ($LiNbO_3$), lithium tantalite ($LiTaO_3$), potassium titanium oxygenic phosphate ($KTiOP_4$), liquid crystal on silicon (LCOS) or 2D layered material modulator are often utilized to convert electronic signals to the optical domain via generating optical phase carrier or high-order harmonics, which are critical components in modern telecommunication networks and microwave-photonic systems [2, 4].

By inducing the known phase carrier via EOMs to specific optical field, the useful optical information will be coded/manipulated, and frequency character will be re-distributed, after receiving optical field by photodetectors or spectrometers, the delivered optical information could be reconstructed through algorithms [5–10]. In practical opto-electronic applications, the extensively-studied and applied $LiNbO_3$ often employs high driven voltage (>100 V) for relatively-low frequency band (DC to 1 GHz) in bulky size, while the resonant $LiNbO_3$ allows a radio-frequency slewing

IntechOpen

rate (>1 GHz) at low driven voltages (<50 V, compatible to the CMOS devices). In addition of constant direct voltage, the external-applied electrical waveform could be sinusoidal, triangle, saw-tooth, trapezoid or other artificial waveforms.

Methods of self-growth of original bulky lithium niobate crystal is mature now, the real challenging mostly focuses on: (1) How to micro-miniaturize EOMs, massively-integrate them on various substrates [11–14]. (2) How to give large phase or amplitude variation in a tiny size EOMs, where short optical path inside EOM is helpful to decrease optical propagation loss [12, 15–17]. (3) How to improve their electro-optical efficiencies/group velocity matching/optical damage threshold [14, 17]. The solution of above challenges is highly-valued in optical communication networks/microwave photonics, quantum photonics and non-reciprocal optics [18].

Of great interest to broad application, the technological advance of micro/nanofabrication, high-power ultrafast laser processing makes the stiff, transparent crystal be artificially fabricated with ultrafine spatial resolution on substrate via high instantaneous power nanosecond or picosecond laser, which provides the key to integrate EOMs. Predictably, the imminent era of on-chip photonic devices via integrated EOMs will revolve optics, microwave, telecommunication, large-volume memory by providing tremendous new photonic applications.

2. Applications of lithium niobate

$LiNbO_3$ is a trigonal structure, wide transmittance range, non-linear optic crystal. Due to the spontaneous polarization, there exist electric dipole moments in lithium niobate crystals. The direction of these electric dipole moments tends to be consistent with the direction of outside-applied electric field, which affects the refractive index of the crystals. When being injected alternative or constant electric field, the resultant refractive index changes accordingly, making the optical phase shift as designed (seen in **Figure 1**), which is known as electro-optic effect. In optical communication or photonics, the optical signal passes through the lithium niobate crystal waveguide with alternating voltage, its equivalent optical path difference changes, inducing a controllable periodical phase shift or amplitude fluctuation [19].

Inside EOMs, the varying refractive index Δn presents a linear relationship with the applied electric field E:

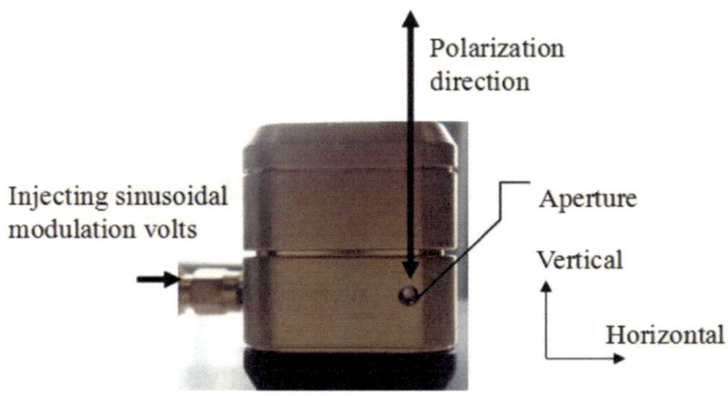

Figure 1.
A box-sealed, commercial magnesium oxide (MgO)-doped electro-optic modulator (4002, Newfocus), the crystal is installed with electronic drivers, diameter of aperture is <2 mm, allowing light to travel, phase-shifting happens only in the polarization direction.

$$\Delta n = \left(\frac{1}{2} n_e^{\,3} r_{33}\right) E \tag{1}$$

Subsequently, the optical phase shifting replies on length and thickness of EOM and the outside-applied voltage V:

$$\phi_m = \frac{2\pi}{\lambda}\left(\frac{1}{2} n_e^{\,3} r_{33}\right)\frac{d}{h} V \tag{2}$$

h is thickness, ϕ_m is the generated phase shift, n_e is dynamic refraction index of EOM injected with electric signal, λ is wavelength of light, r_{33} is the intrinsic nonlinear coefficient of EOM. The optical power damage threshold of MgO-doped EOMs (doping concentration: 5%) greatly increases, for general laser sources, the possibility of laser burning of electro-optic crystal becomes very small, which is beneficial to the long-term operation incorporating with high-power laser beam.

To get better electro-optic modulation efficiency, the polarization of laser and EOM should be paralleling. Otherwise, the dis-match will decrease the induced phase or amplitude shifting. It is noteworthy, the EOMs having intrinsic polarization could be furtherly used in polarization-multiplexing optical system, which manipulate light at different polarization without optical cross-talk (**Figures 2** and **3**).

As simple demonstrated in **Figure 2**, the emitted laser beam is reflected by external vibrating target and injects into laser cavity, the reflected light affect by optical Doppler effect carries the information of external target to create a new fluctuated optical signal. When this optical signal travel in and out an EOM, the information is decoded artificially, which is a simple typical optical modulation application. By introducing polarization-multiplexing technique (where light has at least two polarizations), EOMs could modulate lights at different polarization for multi-channel communication, computation or sensing (as exampled by **Figure 3**) [9].

With technological advance, more situations (in-fiber optical communication) requires embedded in-fiber EOMs instead of spatially box-sealed EOMs, and highly-integrated photonic devices requires on-chip arrays of EOMs. Therefore, in order to minimize the weight and reduce the volume of $LiNbO_3$ for in-fiber embedded optical systems, $LiNbO_3$ wafer (or chip, widely fabricated via "ion slicing and direct bonding," seen in **Figure 4**) is optical lithographic patterned, cut and then annealing proton exchanged to form Y-type/S-type/line waveguide shapes. After polishing, the polished waveguide is aligned with polarization-maintaining fiber,

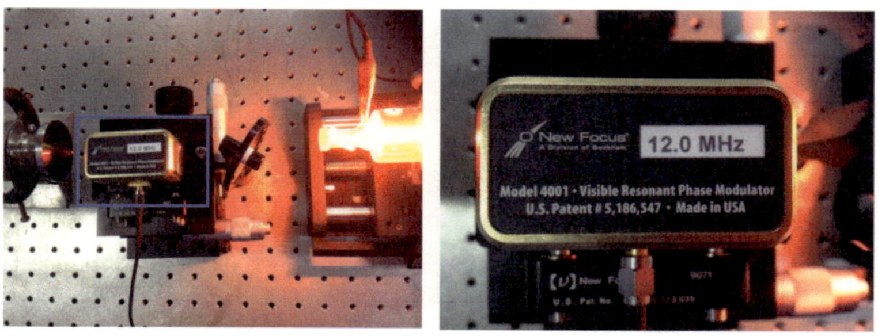

Figure 2.
A simple feedback laser system for recording sub-ultrasonic vibration of an ultrasound transducer. It consists of four main devices: a linearly polarized He-Ne laser (model: JW1, Beijing Topda laser), adjustable gradient optical density attenuator (model: NDC-50C-2M-A, Thorlabs), resonant phase modulator (model: 4001nf, Newfocus, 12 MHz central frequency, 36 V safe operating voltage, which is sealed with a resonant electric circuit) and an ultrasound transducer [10].

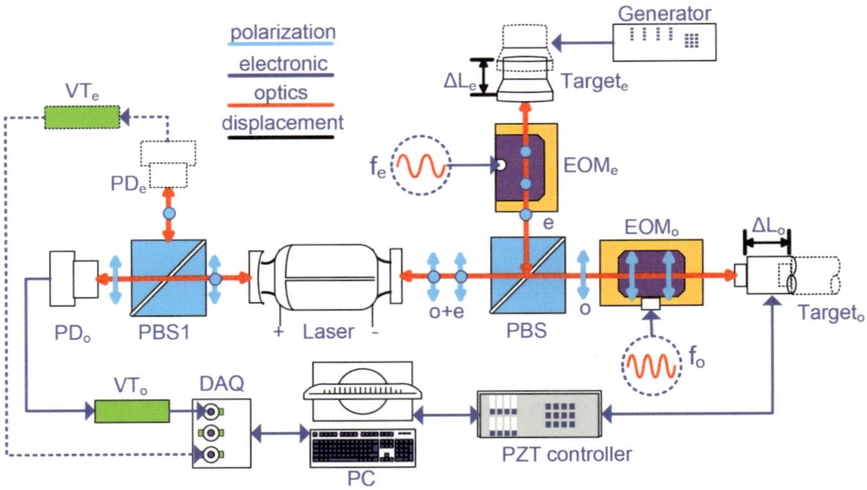

Figure 3.
A lensless two-dimensional polarization multiplexing phase-shifted self-mixing interferometry system is presented. The main devices include a two-longitudinal mode He-Ne laser, two electro-optic crystals with perpendicular polarizations, polarization splitter prism and two independent measured objects in different directions. Two linearly polarized lights are defined as o and e, VT represents adjustable resistance, computer (PC) takes responsibility of signal processing. The e-channel source is a speaker driven by AC voltage, and the o-channel source is a closed-loop one-dimensional piezoelectric ceramic (PZT). In this polarization-multiplexing system, the formed optical intensity fluctuations induced by the measured two external targets are modulated at different-frequency phase-shifting, therefore, no overlapping error happens after electro-optic modulation/manipulation [9, 10, 20].

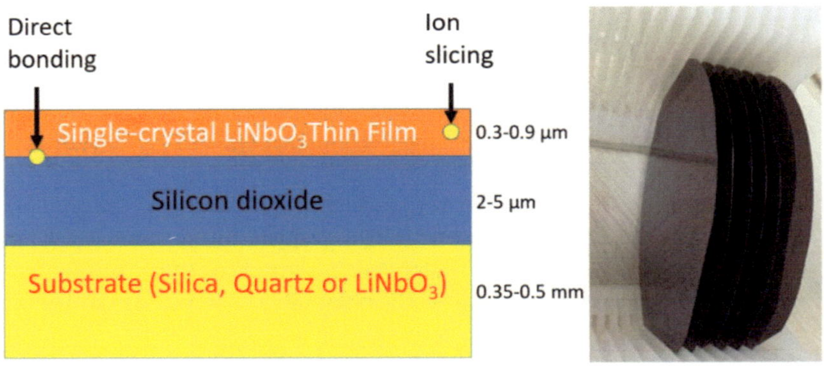

Figure 4.
The standard "ion slicing and direct bonding" process for fabricating homogeneous or heterogeneous substrate-based thin $LiTaO_3$ or $LiNbO_3$ film, the right side are the fabricated $LiNbO_3$ wafers, which could be further X-cut or secondary etched for chip-scale electro-optic devices.

and then fixed in the shell. After connecting the lead wire of the electrode, the polished surface is packaged as in-fiber waveguide phase (or amplitude) modulator. Better than box-sealed EOMs, the integrated Y-shape waveguide phase modulator is easy to couple with in-fiber system and becomes bendable and suitable for fiber optical communication [13].

Y-shaped branch $LiNbO_3$ phase modulator contains three segments: input straight waveguide, symmetrical two S-shaped curved waveguide and two output straight waves. In the guide section, the input straight waveguide undertakes the tasks of connecting with the input optical fibers and equally distributing the power when the light enters the waveguide, and the two after the light comes back through the optical fiber. The minimum distance between two output waveguides must be

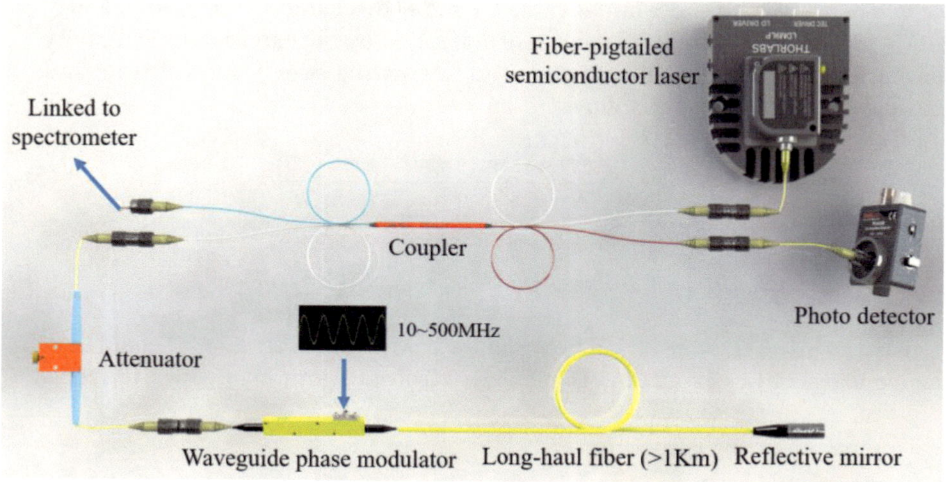

Figure 5.
A wave-guide phase modulator is inserted to the flexible, long-haul, in-fiber interferometric sensing system. Waveguide electro-optics modulator (Photoline, NIR-MPX800-LN-0.1, WPM) is driven by a low-voltage electronic signal (actually, using a microwave signal generator, the modulation frequency could be far beyond 500 MHz). The photo-electric detector (PD) with flange plate records the front output of semiconductor laser diode for further digital analysis, and an in-built photodetector inside semiconductor laser records the another optical intensity.

>0.25 mm when the diameter of polarization maintaining fiber (with cladding) is 0.25 mm. In the package process, polarization-maintaining fibers need to be bonded on small blocks before bonding with chips.

As demonstration of an in-fiber Y-shape phase modulator in laser beam manipulation, **Figure 5** illustrates a bendable optical configuration using in-fiber sealed waveguide type EOM, where fiber-pigtailed semiconductor laser (SL) is linked to a fiber-coupler, an optical power attenuator, and a 1 km-length standard single mode fiber (SMF). The Fabry-Perot SL (Thorlabs, LP852-SF30) emits an invisible 850 nm light, and possesses a sufficient sensitivity to the back-scattered weak light. A fraction of the emitted infrared light is back-scattered by an in-fiber mirror to re-enter into SL. To obtain optical phase variation, herein, the high slewing rate waveguide phase modulator (WPM) resonates from 10 to 500 MHz for phase shifting according to electric driven signal. Therefore, gauss beam of SL was transformed to a Bessel beam with over 10 M optical subdivisions.

Subsequently, logarithmic optical power loss of the 1180 m-length SMF (Yangtze, C5B00419AB2) by an optical power meter (PM100D, Thorlabs) is only 2.68 dB/km attenuation at 850 nm wavelength. The small power loss and high photon-current conversion efficiency of the SL guaranteed a long haul sensing. Semiconductor laser oscillation during the interferometric process depends on the reflected light and the phase shifting by the microwave-band WPM, which strictly follows the classical light-speed equation.

When the emitted and reflected optical fields couples together, and triggers laser frequency shifting $f \rightarrow f_0 + \Delta f$, the governing equation of dynamical oscillation condition changes to this form:

$$(g_0 + \rho \Delta N) f \frac{4\pi l}{c} = -\eta \cos\left(4\pi f \frac{\varphi}{c}\right) \tag{3}$$

$$(n_0 + \chi \Delta N) f \frac{4\pi l}{c} + \eta \sin\left(4\pi f \frac{\varphi}{c}\right) = 2\pi \tag{4}$$

where n_0 and g_0 denote initial values. χ and ρ are changing speeds of n and g to ΔN respectively. The delivered phase is affected by the external optical cavity length. The shifting optical frequency and fluctuating laser powers of front and rear facets are expressed as follows:

$$\Delta f = \frac{-\eta\sqrt{1+\alpha^2}\,\sin\left(4\pi f\varphi/c - \arctan(\alpha)\right)}{4\pi n_0 l\left(1-\frac{g_0}{n_0}\alpha\right)} \tag{5}$$

$$\Delta P_{front/rear} = \mp m P_0 \cos\left(\varphi + \frac{1}{2}n_e^3 r_{33} E \sin\left(2\pi f_{RF} t\right) + \varphi_{speckle}\right) \tag{6}$$

where $\alpha = \frac{\chi}{\rho}$, E is the electrical field applied on the WPM. The Eq. (6) could be demodulated as Bessel functions using phase-resolved algorithm. Because the added

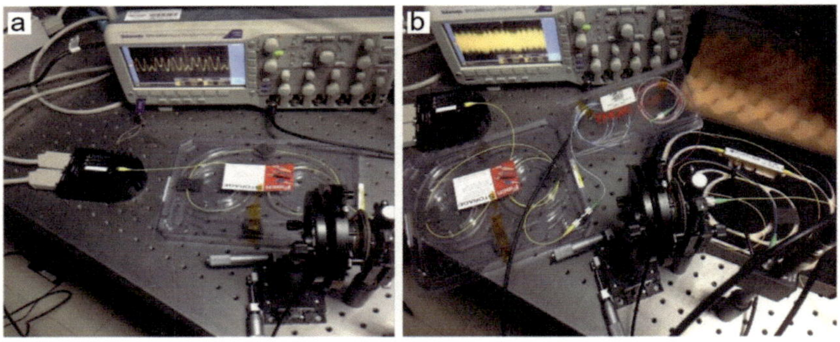

Figure 6.
Photos of in-fiber optical sensing experimental setup and observation of optical intensity fluctuation via oscilloscope. (a) Top view of vibration measurement without optical fiber phase modulator. (b) The view after inserting waveguide phase modulation is presented, including two-end input and two-end output optical fiber couplers, optical fiber waveguide phase modulator, loudspeakers as external reflector, the modulator is driven by a sinusoidal 17.5 MHz, 5 V electric voltage.

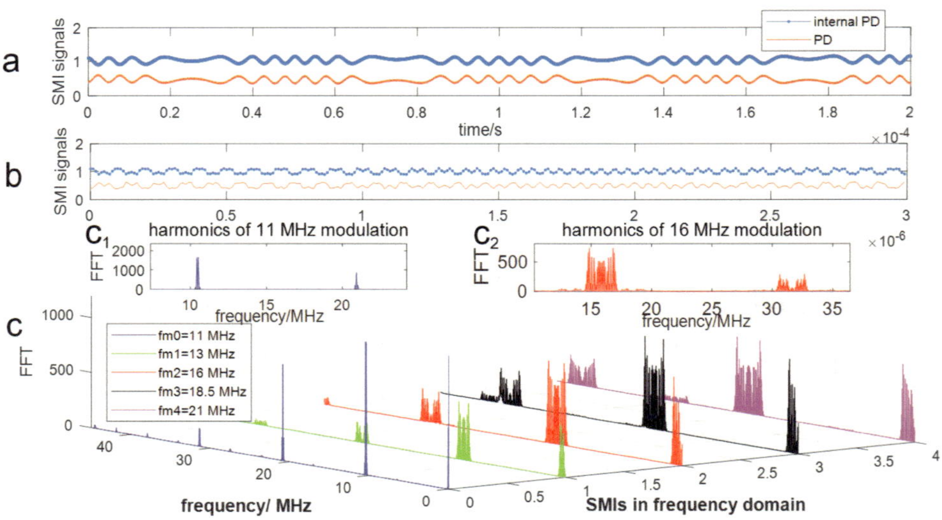

Figure 7.
Interferometric optical signals in time and frequency domains. (a) Synchronously-detected interferometric singles without activating electro-optics modulation (from PD and the photodetector of SL), presenting phase reversal phenomenon. (b) Interferometric signals of identical phase reversal when WPM slewed at 10 MHz/s. (c) Phase shifted interferometric signals in frequency domain. Zoomed-in figures C_1 and C_2 showed first-order and second-order harmonics, respectively.

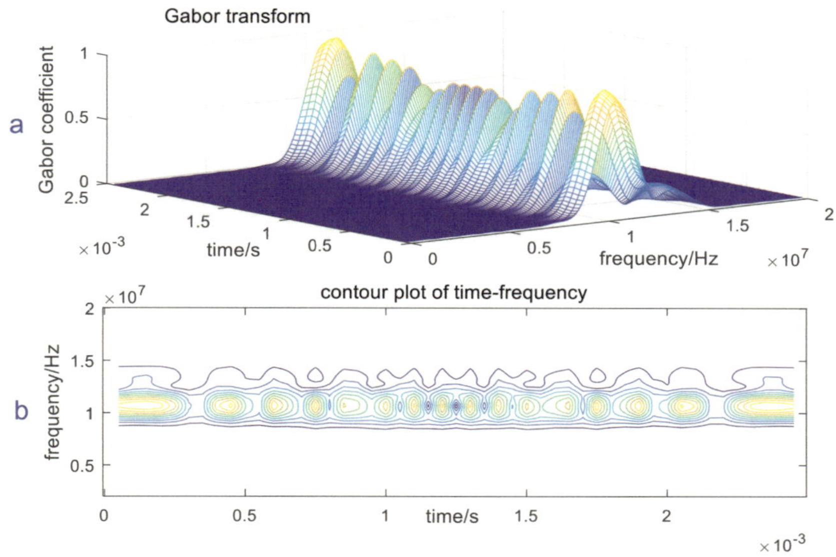

Figure 8.
The first harmonic in joint time-frequency domain transformed by Gabor window when the modulation frequency was 11 MHz. (a) Three-dimensional distribution of Gabor coefficient of first harmonic of the phase-shifted optical field. (b) Time-to-frequency contour plot of (a), the center frequency fluctuates at around 11 MHz, known as the frequency of the applied electric field.

phase shifting is much faster than the target phase information, phase shifting induced by WPM operates as high-frequency phase carrier in subsequent decoding (**Figure 6**).

The frequency-domain signals of the electro-optic modulated intensity are obtained from fast Fourier transform (FFT) in **Figure 7**, where the optical harmonics concentrated around phase shifting frequencies with large adjoining space. Sufficient harmonic spaces eliminates the overlapping phenomenon for decoding accuracy. Gabor transform is implemented on the first harmonic of the phase shifted signals to show time-joint-frequency domain characteristics, which is similar to a straight mountain ridge (**Figure 8a**), concluding that the harmonic signal fluctuated around the phase shifting frequency.

3. Fabrication and integration

Currently, surface plasmonic polariton (SPP) devices have achieved desirable modulation of phase and amplitude, obtained by mesoscale fabrication technique. However, the metals that give plasmonic such promise are also the largest hindrance, as such devices suffer from large on-state loss. Still, typical insertion losses of 10 dB due to plasmonic propagation remains a concern for high-speed applications. Thus, plasmonic on-chip technologies have been unable to replace the existing electro-optics solutions [12, 15].

$LiNbO_3$-based devices, including in-fiber electro-optics modulators and other frequency converters, which are mostly fabricated via ion diffusion or proton exchange methods, leading to low index contrast and weak optical confinement. To address the future broad application of EOMs in chip-scale [16], the advanced micro/nanofabrication are often employed for sub-wavelength scale, massively-integrated $LiNbO_3$ platform to revolutionize optical communication and microwave photonics [17, 21, 22].

However, due to the hardness and stable chemical properties of $LiNbO_3$, traditional mechanical characterization or chemical etching methods can not achieve the micro/nano structures of lithium niobate with high spatial resolution. This problem

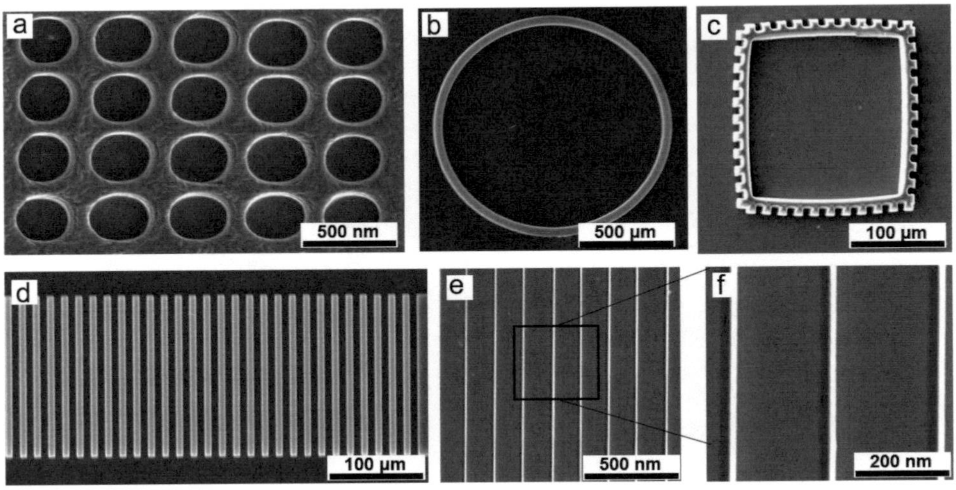

Figure 9.
Scanning electron microscope (SEM) images of (a) the hole array drilled by FIB on $LiNbO_3$ thin film. (b) (c) Ring-shape $LiNbO_3$ waveguide. (c) Square-shape waveguide. (d) Array of micro-wires of $LiNbO_3$ micro-wire circuit. (e) and (f) Zoomed-in SEM image of nanowires.

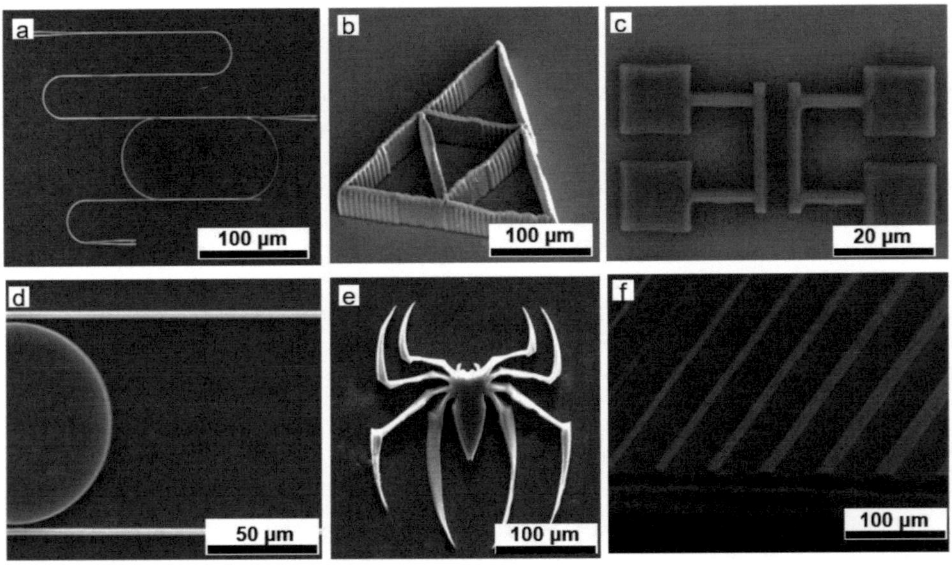

Figure 10.
(a–f) More fabricated photonic $LiNbO_3$ waveguides and patterns on silicon, which demonstrate ultra-fine and flexible features.

has greatly hindered the development of miniaturized and integrated lithium niobate photoelectric chips and devices. Major roadblock for the practical application of integrated $LiNbO_3$ photonics is the difficulty of fabricating devices that simultaneously achieve low optical propagation loss and high confinement.

Recently-developed thin-film $LiNbO_3$-on-insulator technology makes this possible. Several mainstream techniques like the acid etching, ion etching, plasma etching, optical lithography, high-power nanosecond or picosecond laser writing, electron beam writing or focused ion beam (FIB) etching are used to fabricate and integrate the tiny-size $LiNbO_3$ (nano-scale surface roughness and complicated arrangement, usually on micro-scale silicon dioxide or silicon substrates) waveguide for device-level applications. The fabrication resolution of ultrafast laser cutting is typically in the order of 1 μm for glass, semiconductors and metals. However, today's optical lithography can easily achieve sub-micron resolutions [15, 18, 23, 24].

The standard electron beam lithography is firstly used to define patterns in photoresist, the patterns are subsequently transferred into the $LiNbO_3$ thin film using plasma reactive ion etching, the plasma power and chamber condition are artificially tuned to remove $LiNbO_3$ or contaminations, and after removing the photoresist, the pre-designed $LiNbO_3$ patterns in wafer/chips will be obtained. Or, by selectively bombarded and removed lithium niobate molecule via high-energy gallium ion beam, the arrays of nanowires, or nanowall-like $LiNbO_3$ devices could be obtained (seen in scanning electron microscope, SEM images, in **Figures 9** and **10**). Moreover, the nanofabrication technique could provide optical micro-resonators or photonic crystals. After furtherly interconnecting with driven electric circuits, on-chip integrated EOMs are expected to provide huge controllable photonic functions.

Photonic circuits based on $LiNbO_3$ waveguide have shown the potential for complex information processing systems employing both quantum and classical light sources [24]. To increase computational efficiency, new substrate materials (silicon, semiconductor materials, or fused silica) have been utilized to construct complicated resonant electro-optics modulators [18]. Silicon provides high refractive index enabling the fabrication of photonic circuits with strong confinement (seen in **Figure 10**). Specifically, silicon photonics maximize their modulation depth by using the resonant structures, enabling more compact components than in-fiber integration [17].

4. Conclusions

The electro-optics modulation capability of $LiNbO_3$, box-sealed packaging, applications of interferometric optical sensing via electro-optic manipulating light field are firstly demonstrated here. Then, the in-fiber integration of electro-optic modulation, and the existing micro/nano fabrications for on-chip $LiNbO_3$ waveguides are introduced. The background knowledge of $LiNbO_3$, advantages and inferiority of different kinds of sealing method and fabrication methods are presented as reminder for near-future broad applications of EOMs. We believe the integrated EOMs, is the future of the emerging on-chip microwave photonics, quantum optics and optical communication network. The ultralow loss, high optical confinement and the ability to connect to microwave electrodes will bring electro-optic and nonlinear optical systems to a micro/nano scale, massive application that has been inaccessible until now, which deserves a better and deeper investigation.

Acknowledgements

This chapter is supported by the National Science Youth Fund of china (61805094), China Postdoctoral Science Foundation (2017M622417). Author want to thank Nanjing Normal University for providing experimental setup, and thank the facility support of the Center for Nanoscale Characterization & Devices (CNCD), WNLO of HUST for providing SEM images.

Conflict of interest

The author declare no conflict of interest in this chapter.

Author details

Yufeng Tao
Wuhan National Laboratory for Optoelectronics, Huazhong University of Science and Technology, Wuhan, China

*Address all correspondence to: wsnwp520@sina.com

References

[1] Oulton RF, Sorger VJ, Genov DA, Pile DFP, Zhang X. A hybrid plasmonic waveguide for subwavelength confinement and long-range propagation. Nature Photonics. 2008;**2**:496-500. DOI: 10.1038/nphoton.2008.131

[2] Gheorma IL, Osgood RM. Fundamental limitations of optical resonator based high-speed EO modulators. IEEE Photonics Technology Letters. 2002;**14**:795-797. DOI: 10.1109/lpt.2002.1003096

[3] Haffner C et al. Plasmonic organic hybrid modulators: Scaling highest speed photonics to the microscale. Proceedings of the IEEE. 2016;**104**: 2362-2379. DOI: 10.1109/jproc.2016.2547990

[4] Sun ZP, Martinez A, Wang F. Optical modulators with 2D layered materials. Nature Photonics. 2016;**10**:227-238. DOI: 10.1038/nphoton.2016.15

[5] Alloatti L, Cheian D, Ram RJ. High-speed modulator with interleaved junctions in zero-change CMOS photonics. Applied Physics Letters. 2016;**108**:131101. DOI: 10.1063/1.4944999

[6] Maier SA et al. Plasmonics—A route to nanoscale optical devices. Advanced Materials. 2001;**13**:1501-1505. DOI: 10.1002/adma.200390134

[7] Vahala KJ. Optical microcavities. Nature. 2003;**424**:839-846. DOI: 10.29172/5802dce1007946889f4ea79871191efc

[8] Xu Q, Schmidt B, Pradhan S, Lipson M. Micrometre-scale silicon electro-optic modulator. Nature. 2005;**435**:325-327. DOI: 10.1038/nature03569

[9] Tao YF, Wang M, et al. Compound cavity theory of resonant phase modulation in laser self-mixing ultrasonic vibration measurement. Optical Engineering. 2016;**55**:074107. DOI: 10.1117/1.oe.55.7.074107

[10] Tao YF, Wang M, et al. Carrier-separating demodulation of phase shifting self-mixing interferometry. Optics and Laser Technology. 2017;**89**:75-85. DOI: 10.1016/j.optlastec.2016.08.013

[11] Poberaj G, Hu H, Sohler W, Günter P. Lithium niobate on insulator (LNOI) for micro-photonic devices. Laser & Photonics Reviews. 2012;**6**: 488-503. DOI: 10.1002/lpor.201100035

[12] Wang C, Zhang M, Stern B, Lipson M, Lončar M. Nanophotonic lithium niobate electro-optic modulators. Optics Express. 2018;**26**:1547-1555. DOI: 10.1364/oe.26.001547

[13] Mercante AJ et al. 110 GHz CMOS compatible thin film $LiNbO_3$ modulator on silicon. Optics Express. 2016;**24**:15590-15595. DOI: 10.1364/oe.24.015590

[14] Jin S, Xu L, Zhang H, Li Y. $LiNbO_3$ thin-film modulators using silicon nitride surface ridge waveguides. IEEE Photonics Technology Letters. 2016;**28**:736-739. DOI: 10.1109/lpt.2015.2507136

[15] Haffner C et al. Low-loss plasmon-assisted electro-optic modulator. Nature. 2018;**556**:483-486. DOI: 10.1038/s41586-018-0031-4

[16] Sun C et al. Single-chip microprocessor that communicates directly using light. Nature. 2015;**528**:534-538. DOI: 10.1038/nature16454

[17] Reed GT, Mashanovich G, Gardes FY, Thomson DJ. Silicon optical modulators.

Nature Photonics. 2010;**4**:518-526. DOI: 10.1038/nphoton.2010.179

[18] Janner D, Tulli D, García-Granda M, Belmonte M, Pruneri V. Micro-structured integrated electro-optic $LiNbO_3$ modulators. Laser & Photonics Reviews. 2009;**3**:301-313. DOI: 10.1002/lpor.200810073

[19] Tao YF. Key technology of self-mixing micro/nano vibration measurement [thesis]. Nanjing Normal University; 2017

[20] Tao YF, Xia W, et al. Integration of polarization-multiplexing and phase-shifting in nanometric two dimensional self-mixing measurement. Optics Express. 2017;**25**:2285-2298. DOI: 10.1364/oe.25.002285

[21] Fortier TM et al. Generation of ultrastable microwaves via optical frequency division. Nature Photonics. 2011;**5**:425-429. DOI: 10.1038/nphoton.2011.121

[22] Tzuang LD, Fang K, Nussenzveig P, Fan S, Lipson M. Non-reciprocal phase shift induced by an effective magnetic flux for light. Nature Photonics. 2014;**8**:701-705. DOI: 10.1038/nphoton.2014.177

[23] Tao YF, Wei CYR, Liu JW, Deng CS, Cai S, Xiong W. Nanostructured electrically conductive hydrogels obtained via ultrafast laser processing and self-assembly. Nanoscale. 2019;**11**(18):9176-9184. DOI: 10.1039/C9NR01230C

[24] Wooten EL et al. A review of lithium niobate modulators for fiber-optic communications systems. IEEE Journal of Selected Topics in Quantum Electronics. 2000;**6**:69-82. DOI: 10.1109/2944.826874

Chapter 4

Polarization Modulation

Ichihiko Toyoda

Abstract

Conventional wireless communication systems use amplitude, frequency, and phase of the carrier wave to carry information. However, actual radio waves also have vector parameters, such as polarization and propagation direction. In this chapter, a modulation scheme using polarizations is explained. The polarization modulation provides an additional degree of freedom for the modulation of the carrier waves. Furthermore, the polarization modulation is suitable to realize simple transceivers using RF signal processing. Antennas are the most important key parts of the polarization modulation systems. Polarization agile antennas, active integrated array antenna which integrates an oscillator and modulators, and polarization discrimination antenna are also introduced.

Keywords: polarization, antenna, polarization modulation communication, polarization discrimination, active integrated array antenna, RF signal processing

1. Introduction

Nowadays, consumer wireless communication systems like a cell phone and wireless local area network (WLAN) are widely used in the world, and they have become an indispensable part of our daily lives. Much higher data rate and higher capacity are strongly required as well as high-frequency utilization efficiency. To meet the requirements, many advanced technologies such as multiple-input, multiple-output (MIMO) and orthogonal frequency-division multiple access (OFDMA) have been developed.

In the classical wireless communication systems, only the time domain parameters such as amplitude, frequency, and phase are used to modulate the carrier wave. However, actual radio waves are vector quantity, and they have spatial parameters like a polarization and direction of propagation. In the next-generation wireless communication systems, utilization of the spatial parameters is essential to achieve higher data rates, larger capacity, and higher-frequency utilization efficiency since the spatial parameters have not been effectively used in traditional wireless communication systems. Recent advanced wireless communication systems utilize a part of these spatial parameters. MIMO and polarimetric radar as well as the traditional polarization diversity are the examples. The massive MIMO technology, which is expected to be employed in the fifth-generation (5G) mobile communication system, is another example. However, these systems require power-consuming digital signal processing. Therefore, RF signal processing technology utilizing the characteristics of the radio wave is expected to realize advanced transceiver module for many wireless applications. To achieve wireless communication systems which effectively utilize the spatial parameters, antenna technology based on the RF signal processing is one of the most important technologies.

In this chapter, the basic concept of the wireless communication system employing a polarization modulation scheme and antenna technology based on the RF signal processing is introduced.

2. Polarization modulation communication

As classic wireless communication systems use the amplitude, frequency, and phase of the carrier wave to carry information, the radio wave is treated as a scalar signal as follows:

$$s(t) = A\sin(2\pi ft + \varphi) \tag{1}$$

where A, f, and φ are the amplitude, frequency, and phase of the carrier wave, respectively. However, the actual radio wave is a vector signal. For example, the electric field of the radio wave is given by

$$\boldsymbol{E}(\boldsymbol{r}, t) = \boldsymbol{E}_0 \sin(2\pi ft - \boldsymbol{k}\cdot\boldsymbol{r} + \varphi) \tag{2}$$

where $\boldsymbol{E}_0$ is the vector amplitude that shows the direction of the electric field, i.e., the polarization, and $\boldsymbol{k}$ is the wave number vector specifying the direction of propagation. Even though the spatial parameters have not been effectively used in classic wireless communication systems, these spatial vector parameters have the potential to realize new wireless systems.

Figure 1 shows the basic concept of the wireless communications using the polarization modulation. The transmitter (TX) antenna radiates radio wave while changing its polarizations between $+45^\circ$ and -45° according to the input binary data. At the receiver (RX), the polarization of the radio wave is detected, and the binary data are recovered. As the binary data can be transferred using the orthogonal polarizations as shown in this figure, the polarization can be used as an additional modulation parameter.

Figure 2 shows a vector diagram of the polarization modulated signal. The $\pm45^\circ$ polarizations can be decomposed into the x and y components as shown by the red and blue arrows, respectively. As only the y component is changed according to the data, the $\pm45^\circ$ polarization modulated signal is equivalent to the composition of a binary phase shift keying (BPSK) signal and carrier wave. Therefore, the polarization modulated signals can be generated by simply inverting one of the orthogonal polarizations. As the phase inversion of the signal is easily achieved in RF, the polarization modulation scheme is suitable for RF signal processing and realizing a simple transmitter. The polarization modulation also provides great benefits to receivers. As the carrier wave component is included in the polarization modulated signal and it is transmitted to the receiver, the receiver does not require

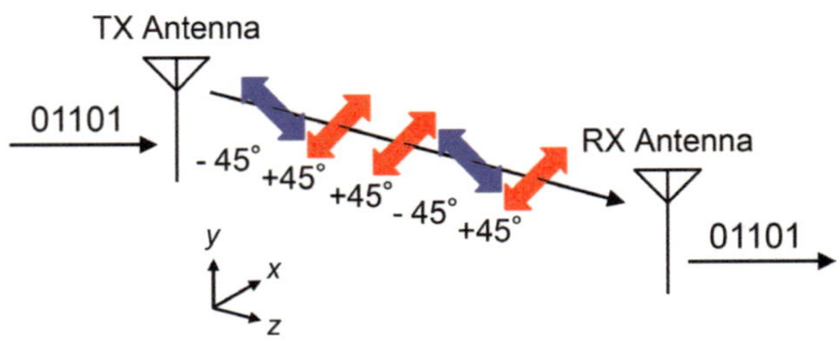

Figure 1.
Basic concept of the polarization modulation communications.

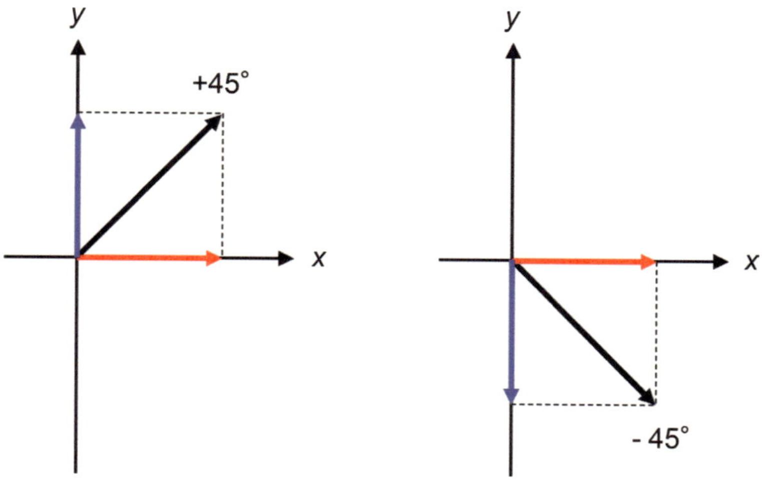

Figure 2.
Vector diagram of the polarization modulation.

carrier recovery circuits, and the phase noise of the local oscillator does not affect the performance of the communication system. The detection in RF with simple receiver configuration is also applicable.

Antennas are one of the key elements to achieve polarization modulation systems because polarizations are generated in antennas. The antennas used in the polarization modulation systems have to switch their polarizations according to the input data. Therefore, polarization agile antennas are required [1, 2].

3. Basics of polarization agile antennas

Basically, any polarizations can be expressed in the sum of two orthogonal polarizations. For example, the electric field propagating along the z-axis can be expressed as follows:

$$\boldsymbol{E}(z,t) = \boldsymbol{i}_x E_x \sin(2\pi ft - kz) + \boldsymbol{i}_y E_y \sin(2\pi ft - kz + \varphi) \tag{3}$$

where $\boldsymbol{i}_x$, $\boldsymbol{i}_y$, E_x, E_y, and φ are the unit vectors of the x and y direction, the amplitude of x and y component of the electric field, and the phase difference between the components, respectively. Eq. (3) expresses elliptical circularly polarized waves in general. When the electric field has only the x or y component, the radio wave becomes a linearly polarized wave. When $\varphi = 0$ or π, the radio wave also becomes a linearly polarized wave. On the other hand, the radio wave becomes a circularly polarized wave when $\varphi = \pm\pi/2$ and $E_x = E_y$.

Figure 3 shows basic configurations of several types of polarization agile antennas. The antenna shown in **Figure 3a** is a linear polarization switchable antenna, and it consists of a switch and dual-polarized antenna which radiates horizontal and vertical polarizations. The polarization modulated signals can be excited by simply switching the horizontal and vertical polarizations.

The antenna shown in **Figure 3b** is a circular polarization switchable antenna. A 90-degree hybrid is placed between a switch and dual-polarized antenna. As the input signal fed to one of the input ports of the hybrid is divided into two signals with the phase difference of $\pi/2$, a circularly polarized wave is excited. By switching the input ports of the hybrid, the antenna switches right-handed and left-handed circular polarizations (RHCP and LHCP).

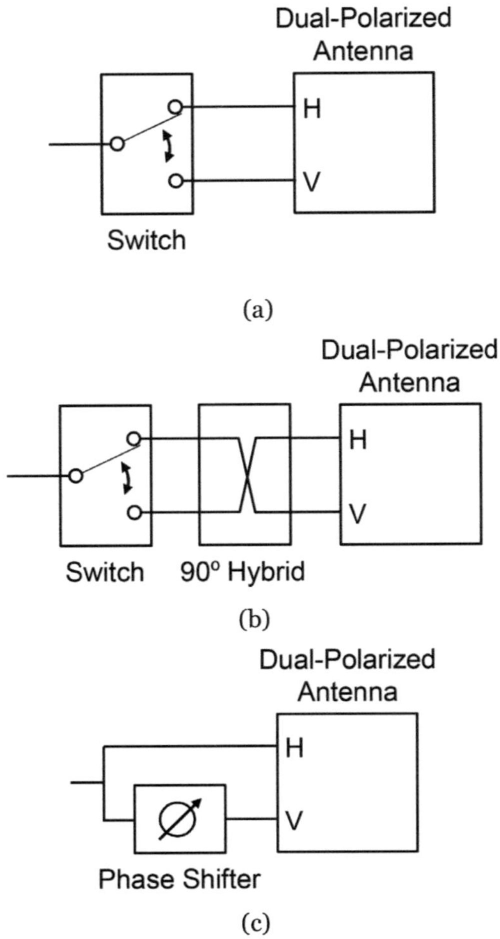

Figure 3.
Basic configurations of polarization agile antenna. (a) Linear polarization switchable antenna; (b) circular polarization switchable antenna and (c) linear/circular polarization switchable antenna.

Figure 3c shows a basic configuration of a polarization agile antenna which switches four polarizations of $\pm 45^{\circ}$ linear polarizations, RHCP and LHCP. The antenna consists of a phase shifter and dual-polarized antenna. When the phase shift value $\varphi = 0$ or π, the antenna excites $\pm 45^{\circ}$ linear polarizations. Furthermore, when $\varphi = \pm\pi/2$, the antenna excites circular polarizations.

4. Practical implementation of polarization agile antenna

In this section, practical implementations of polarization agile antennas employing microstrip antenna elements and planar microwave circuits are introduced.

4.1 Linear polarization switchable antenna

Figure 4 shows a practical implementation of a polarization agile antenna which switches two orthogonal linear polarizations. The configuration is similar to the antenna shown in **Figure 3a**, and it consists of a dual-polarized microstrip array antenna and single-pole double-throw (SPDT) switch [3, 4].

The dual-polarized array antenna has four microstrip antenna elements and employs a feed network using a combination of microstrip lines and slot lines.

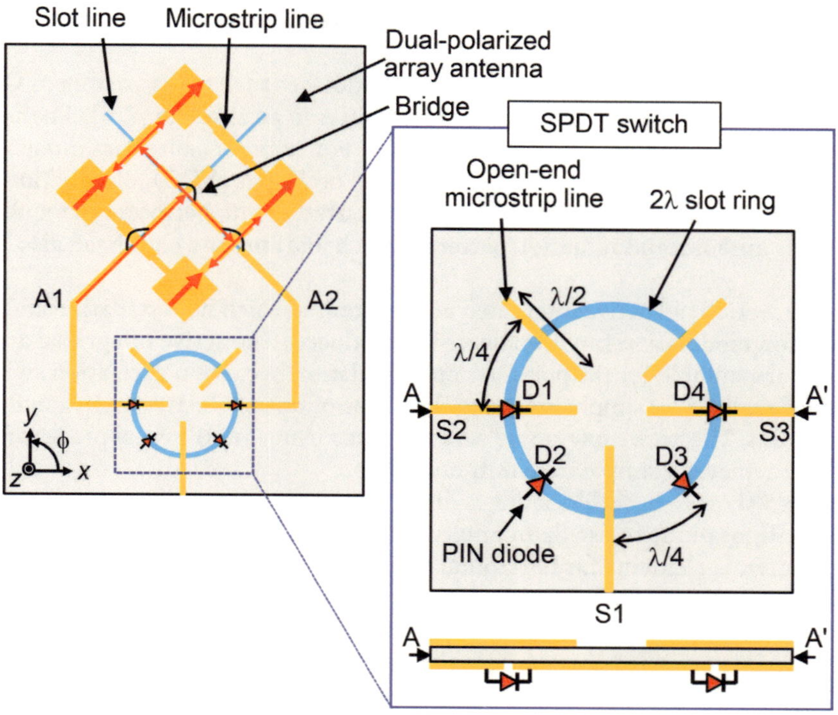

Figure 4.
Structure of a linear polarization switchable antenna [3].

When the signal is fed to the antenna from A1, the signal propagates along the microstrip line as shown by the red line. Here, the signal from A1 is divided into two inphase signals on the slot line. Each signal on the slot line is divided again into two antiphase signals on the microstrip line. Therefore, the signal fed from A1 excites the $+45^{\circ}$-polarized wave as shown by the arrows on the microstrip antenna elements. Similarly, the signal fed from A2 excites the -45°-polarized wave. The input impedance of the array is the same as the input impedance of each microstrip antenna element as the feed network is constructed with a parallel branch and series branch. Therefore, larger array antennas can be easily achieved by simply repeating the same feed structure.

The SPDT switch is constructed with a two-wavelength slot ring, four switching diodes D1–D4, and two half-wavelength open-end microstrip lines. Three microstrip lines are coupled for input and output. The switching diodes are placed over the slot ring with a quarter-wavelength interval. When a positive voltage is applied to the inner conductor of the slot ring, the diodes D1 and D2 become *off*, and D3 and D4 become *on*. Then the signal fed from Port S1 propagates to Port S2 because the *on*-state diodes make short circuits on the slot ring. The half-wavelength open-end microstrip lines also make short circuits on the slot ring. The slot lines from output ports to the open-end microstrip lines act as open circuits because the distance from the diode D1 or D2 to the open-end microstrip line is a quarter wavelength. Similarly, when a negative voltage is applied to the inner conductor of the slot ring, the signal fed from Port S1 emerges at Port S3. Hence, the polarization can be switched by changing the polarity of the voltage applied to the inner conductor of the slot ring.

A circular polarization switchable antenna shown in **Figure 3b** can be realized by placing a 90-degree hybrid between the antenna and switch [5]. A polarization agile antenna which switches four polarizations using phase shifters and magic-T is also demonstrated in [6].

4.2 Active integrated array antenna for polarization modulation

Active integrated antennas integrate active devices such as transistors or Gunn diodes to build in RF signal processing capabilities in an antenna [7, 8]. There are several types of the active integrated antennas. For example, antennas integrating a power amplifier, oscillator, voltage-controlled oscillator (VCO), or injection-locked oscillator have been successfully demonstrated. Furthermore, a frequency-switchable antenna and radiation pattern-switchable antenna have been also proposed.

In this section, an active integrated array antenna which has oscillation and polarization modulation functionalities is introduced. The active integrated array antennas are suitable for the polarization modulation because polarization switching can be realized by simply inverting the phase of one of the two orthogonal polarizations. The active integrated array antenna using an RF signal processing technique achieves a simple transmitter module.

Figure 5 shows a basic block diagram of the active integrated array antenna [9]. In this configuration, an oscillator and two PSK modulators are integrated with two pairs of antenna elements for horizontal and vertical polarization. The oscillator has four output ports and feeds RF signals to the antenna elements. The PSK modulators invert the phase of the RF signals for the vertical polarization. Hence, $\pm 45^{\circ}$ linear polarization switching can be realized.

Figure 6 shows a practical implementation of the active integrated array antenna [10]. A four-port Gunn oscillator with slog-ring resonator is located at the center of the array antenna. Two PSK modulators using a slot ring and PIN diodes are inserted in the feed line for the vertical polarization. The array antenna consists of 12 antenna elements and feed network using microstrip lines and slot lines.

The Gunn oscillator consists of two Gunn diodes mounted on a two-wavelength slot ring. Four microstrip lines are coupled to the resonator with a half-wavelength interval. Therefore, the output ports O1 and O2 (O3 and O4) generate inphase signals, and the phases of O1 and O3 (O2 and O4) become antiphase with each other.

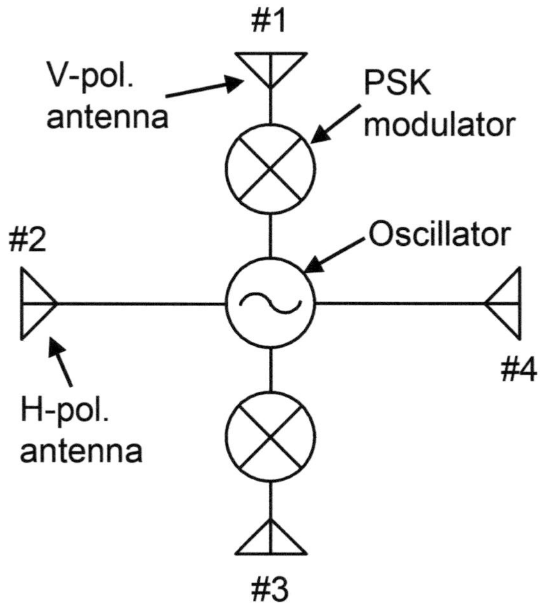

Figure 5.
Basic block diagram of the active integrated array antenna [9].

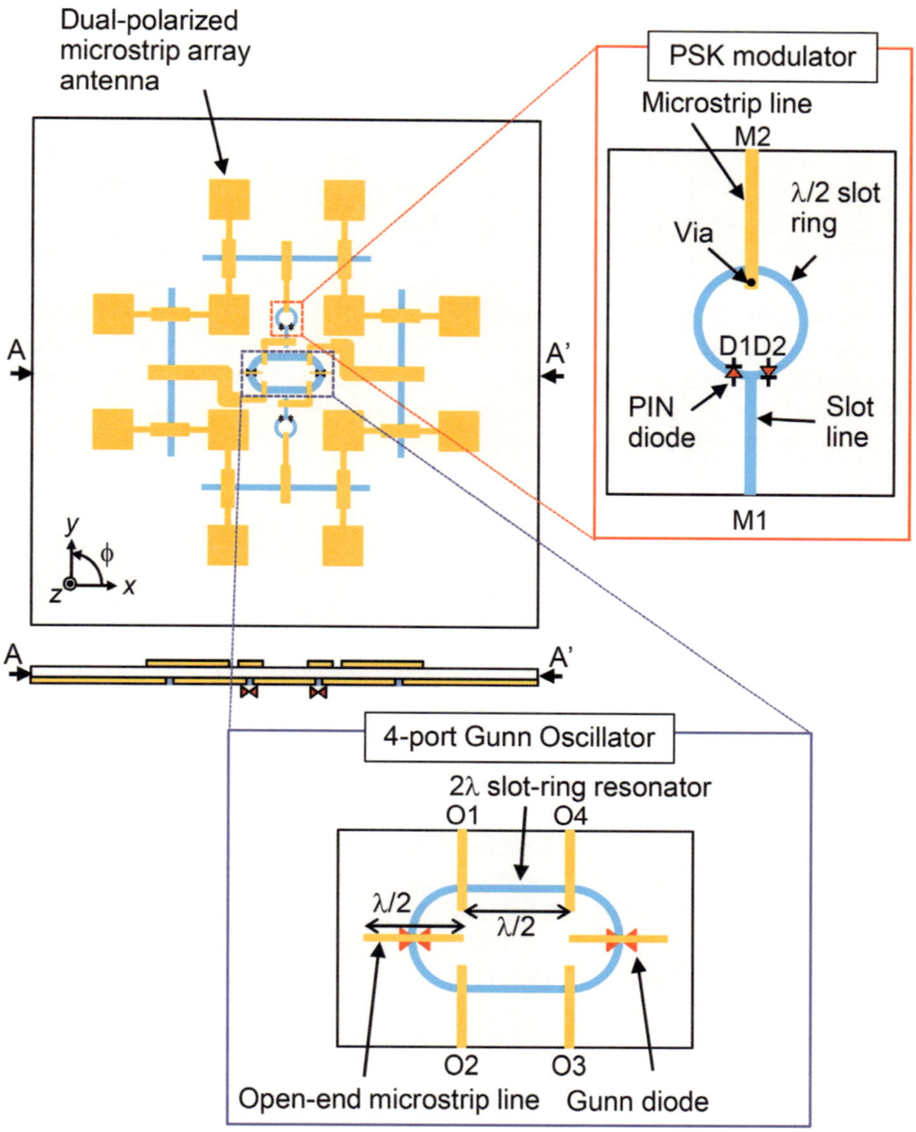

Figure 6.
Structure of the active integrated array antenna [10].

The half-wavelength open-end microstrip lines just above the Gunn diodes stabilize the resonant field in the slot-ring resonator. The bias voltage of the Gunn diodes is applied between the inner and outer conductors of the slot-ring resonator.

The PSK modulator consists of a half-wavelength slot ring and two PIN diodes. A microstrip line and slot line are connected to the slot ring for input and output. The two PIN diodes are mounted at the junction of the slot ring and slot line, and the directions of the PIN diodes are opposite to each other. When positive voltage is applied to the inner conductor of the slot ring, diode D1 becomes *off* and D2 becomes *on*. Therefore, a signal fed to Port M1 propagates along the left half of the slot ring and goes to Port M2. Similarly, when negative voltage is applied, the signal fed to Port M1 propagates along the right half of the slot ring. With this operation, the phase of the signal appeared at Port M2 is inverted by the applied voltage. As a result, polarization switching is achieved.

5. Polarization detection

5.1 Detection of polarization modulated signal

The simplest way to detect the polarization at receivers is to use two orthogonally polarized antennas and compare the signals received by the two antennas.

Figure 7 shows a basic configuration to detect the polarization. Two antennas for horizontal and vertical polarization are connected to a comparator. Comparison of the signals received by the two antennas discriminates the polarizations.

The polarization discrimination can be easily achieved in RF by using an RF multiplier as a comparator. When a polarization modulated radio wave has polarizations of $\pm 45^{\circ}$, the vertical and horizontal components of the radio wave are separately received by using the two orthogonally polarized antennas, and the voltage of each component is expressed as follows:

$$V_H = V \sin(\omega t) \tag{4}$$

$$V_V = V \sin(\omega t + \varphi) \tag{5}$$

where φ is the phase difference between the horizontal and vertical components and it is 0 or π in the case of $\pm 45^{\circ}$ polarization modulation.

The DC output voltage of the multiplier is

$$V_{out} \propto V^2 \cos\varphi. \tag{6}$$

Therefore, when the phase difference $\varphi = 0$, the output voltage of the multiplier V_{out} becomes positive, and it becomes negative when $\varphi = \pi$. Therefore, the polarization modulated signal can be demodulated in RF.

5.2 Polarization discrimination antenna

Figure 8 shows a practical implementation of the polarization discrimination antenna. The antenna consists of 12 microstrip antenna elements, feed network and double-balanced multiplier [11]. The feed network employs microstrip lines and slot lines and achieved simple planar structure. The double-balanced multiplier is located at the center of the array antenna and composed of a slot ring and four detector diodes mounted on the slot ring.

The horizontal and vertical components of the radio wave are separately received by the antenna. The blue and red arrows show the signal of the horizontal and vertical polarizations, respectively. Each received signal is applied to the RF multiplier, and the detected voltage is obtained at the inner conductor of the slot ring.

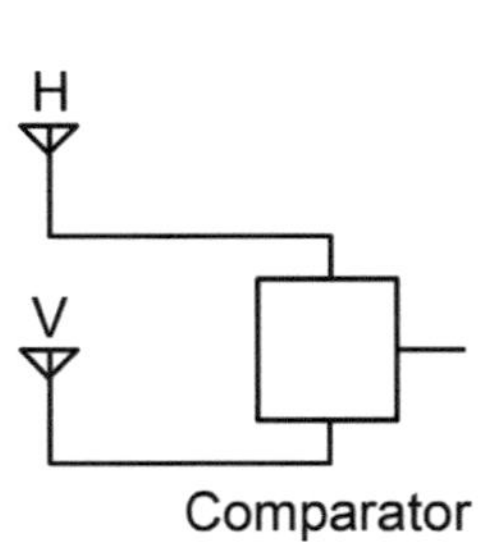

Figure 7.
Basic configuration to detect the polarization.

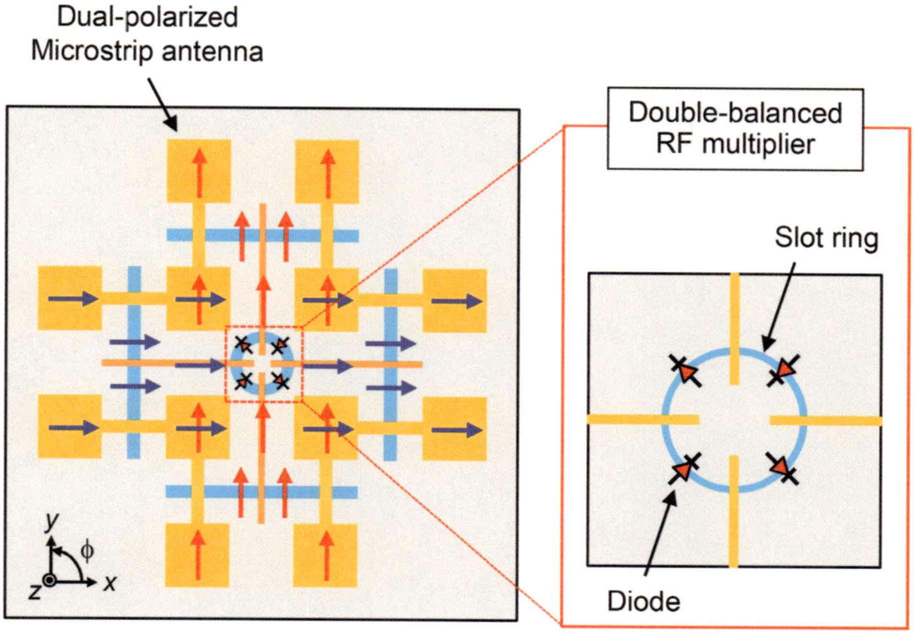

Figure 8.
Polarization discrimination antenna [11].

Polarization discrimination for circular polarizations is similarly achieved by adding 90° phase difference between the horizontal and vertical components [12].

6. Conclusions

In this chapter, a modulation scheme which effectively utilizes the polarization of the radio wave is introduced. The polarization modulation gives a new degree of freedom in the modulation adding to the phase, amplitude, and frequency. Antenna technology is a key to achieve the polarization modulation communication systems. Basics of the polarization modulation and several examples of the polarization agile antennas are introduced. Furthermore, the detection of polarizations and a polarization discrimination antenna are also explained. The concept utilizing polarizations gives new vistas to the next-generation advanced wireless communication systems.

Acknowledgements

The author wishes to appreciate Dr. Eisuke Nishiyama and Dr. Takayuki Tanaka, Associate Professors, Saga University, Japan, for their fruitful discussions. The author also would like to thank Tasuku Uechi of Saga University, Japan, for his technical support and all the students of the Communication Engineering Lab, Saga University, Japan, for their continuous hard work.

This work was supported in part by JSPS KAKENHI Grant Numbers 26420361 and JP17K06429.

Author details

Ichihiko Toyoda
Saga University, Saga, Japan

*Address all correspondence to: toyoda@cc.saga-u.ac.jp

References

[1] Gao S, Sambell A, Zhong SS. Polarization-agile antennas. IEEE Antennas and Propagation Magazine. 2006;**48**:28-37. DOI: 10.1109/MAP.2006.1703396

[2] Haupt RL, Lanagan M. Reconfigurable antennas. IEEE Antennas and Propagation Magazine. 2013;**55**:49-61. DOI: 10.1109/MAP.2013.6474484

[3] Ushijima Y, Nishiyama E, Aikawa M. Single-layer extensible microstrip array antenna integrating SPDT switch circuit for linear polarization switching. IEEE Transactions on Antennas and Propagation. 2012;**60**:5447-5450. DOI: 10.1109/TAP.2012.2207676

[4] Ushijima Y, Nishiyama E, Aikawa M, Toyoda I. Orthogonal linear polarization switchable slot-ring array antenna using RF-MEMS switch. In: Proceedings of 2012 Asia-Pacific Microwave Conference (APMC2012); 3D5-03:809-811; December. 2012. DOI: 10.1109/APMC.2012.6421743

[5] Ushijima Y, Nishiyama E, Toyoda I, Aikawa M. Circular polarization switchable single layer microstrip array antenna. In: Proceedings of the 2012 IEEE International Symposium on Antennas and Propagation (2012 AP-S/USNC-URSI); 409.04; July. 2012. DOI: 10.1109/APS.2012.6348662

[6] Ushijima Y, Nishiyama E, Toyoda I. Polarization agile slot-ring array antenna using magic-T circuit. In: Proceedings of 2012 International Symposium on Antennas and Propagation (ISAP2012); 4A2-3; October. 2012

[7] Qian Y, Itoh T. Progress in active integrated antennas and their applications. IEEE Transactions on Microwave Theory and Techniques. 1998;**46**:1891-1900. DOI: 10.1109/22.734506

[8] Chang K, York RA, Hall PS, Itoh T. Active integrated antennas. IEEE Transactions on Microwave Theory and Techniques. 2002;**50**:937-944. DOI: 10.1109/22.989976

[9] Toyoda I, Furukawa Y, Nishiyama E, Tanaka T, Aikawa M. Polarization agile self-oscillating active integrated antenna for spatial modulation wireless communications. Electronics and Communications in Japan. 2018;**101**:37-44. DOI: 10.1002/ecj.12123

[10] Hasan M, Ushiroda H, Nishiyama E, Toyoda I. A polarization switchable active array antenna integrating a multiport oscillator and PSK modulators. In: Proceedings of 2018 Asia-Pacific Microwave Conference (APMC2018); FR3-B1-2:1253-1255; November. 2018. DOI: 10.23919/APMC.2018.8617433

[11] Hossain MA, Nishiyama E, Aikawa M, Toyoda I. Multi-band orthogonal linear polarization discrimination planar array antenna. Progress in Electromagnetics Research C. 2013;**34**:53-67. DOI: 10.2528/PIERC12080705

[12] Hossain MA, Ushijima Y, Nishiyama E, Toyoda I, Aikawa M. Orthogonal circular polarization detection patch array antenna using double balanced RF multiplier. Progress in Electromagnetic Research C. 2012;**30**:65-80. DOI: 10.2528/PIERC12032402

Chapter 5

Optical Phase-Modulation Techniques

Ramón José Pérez Menéndez

Abstract

Optical phase-modulation technique is a very powerful tool used in a wide variety of high performance photonic systems. Fiber-optic sensors and gyroscopes, integrated-optics sensors, or high-performance photonic integrated circuits are some examples of photonic systems where the optical phase-modulation technique can be efficiently applied. In time, such a photonic system can be integrated as the core part of some specific applications like biosensors, 5G advanced optical communication devices, gyroscopes, or high-performance computation devices. In this work, the main optical phase-modulation techniques are analyzed. Also, a study of the most significant applications of this technique is made, relating it to the most appropriate type in each case.

Keywords: optical phase-modulation, electro-optic phase modulators, sinusoidal phase-modulation, square-wave phase-modulation, triangular phase-modulation, serrodyne phase-modulation, phase-locking technique, phase-locked-loop, optical gyroscopes

1. Introduction

Optical interferometry constitutes an important technique used in a high number of measurement processes for multiple physical magnitudes and quantitative phenomenon [1]. Particularly, fiber-optic waveguides can act as very useful and efficient transmission medium for light guidance in a large group of interferometry-based sensor devices. On the other hand, the study of subject of interferometer fiber-optic sensors has received an extensive treatment in the literature [2, 3]. In this article, a wide analysis of optical phase-modulation is made focused on the optical gyroscopes as the main referenced application. A simple open-loop configuration of interferometric fiber-optic gyroscope based on the Sagnac effect is shown in **Figure 1**. This kind of gyro is based on the Sagnac-effect within an open optical path realized by a N-turn fiber-optic coil when two independent counter-propagating light modes are externally introduced from a broadband laser source through its two ends, respectively. This causes that an interference pattern between the CW and CCW light beams to be collected in a photo-detector with a phase shift given by the following equation, Ref. [4]:

$$\phi_S = \frac{2\pi L D}{\lambda_0 \, c_0} \Omega \tag{1}$$

where L and D are length and diameter of fiber-optic sensing coil, respectively, λ_0 and c_0 are wavelength and speed of light source in vacuum, respectively, and Ω is the rotation rate. **Figure 1** clearly shows that this Interferometer-Fiber-Optic-Gyro (IFOG) has a passive configuration because the laser source is located externally to the sensing coil. In this system, the two counter-propagating light beams travel through the core of a conventional single-mode optical fiber (SMF) by total-internal-reflection phenomenon. As the core diameter of such an optical fiber is only about 8 μm, the spot size of the interference signal can only be coupled to a small area at the end of the fiber loop, for example on the small detection area of a photo-detector. So that, this interference signal affects only one or two interference fringes whose intensity can be evaluated by the following expression:

$$I(\phi) = I_0(1 + \cos\phi) \tag{2}$$

being $I(\phi)$ the output optical signal of interferometer, I_0 the amplitude of each of two CW and CCW counter-propagating beams and ϕ the optical phase-difference between them.

Figure 2 represents the variation of light intensity along a single interference fringe as a function of ϕ. Notice the output intensity noise produced when the phase difference is detected with a phase error $\Delta\phi$. Phase noise sources of this gyro, their influence on output signal and solutions are exhaustively treated in Refs. [5–16].

However, the simple and raw gyro configuration (laser-source, beam-splitter, fiber-coil sensor and photo-detector) showed in **Figure 1** is not effective in practice mainly due to its inability to reduce phase errors. Thus, the decrease in phase error can be effectively achieved by a phase-modulation process of CW and CCW optical

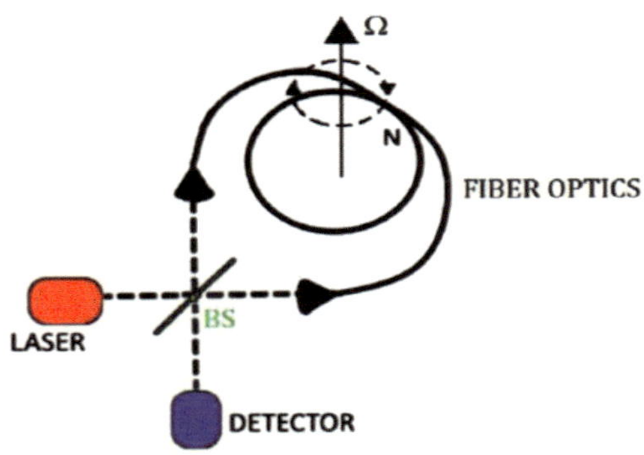

Figure 1.
Basic structure of the IFOG.

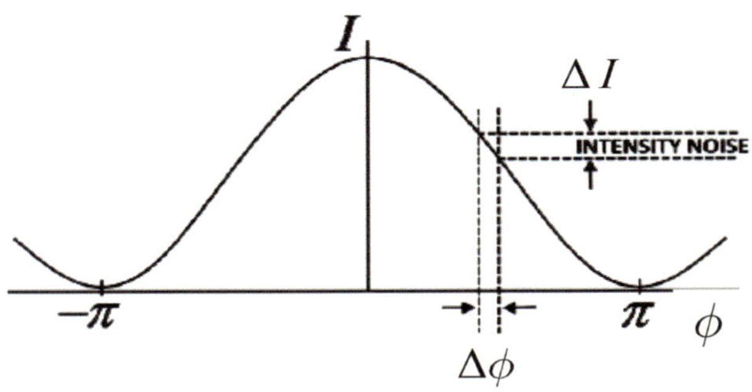

Figure 2.
Two-beam interference response curve as a function of ϕ (phase-difference).

waves by adding an optical phase-modulator in the path of the waves entering the optical fiber sensing coil as it is shown in the arrangement of **Figure 3**. Phase error decreasing is mainly achieved by electronic filtering within phase-sensitive demodulation circuits (PSD).

In this scheme, the immediate consequence of the application of a phase modulation to CW and CCW waves is the need to have an electronic demodulation system for the optical phase (commonly called PSD). This electronic phase-sensitive demodulation system must be located at the electric output of photo-detector. This way, by a demodulation process of electrical output signal of photo-detector, the Sagnac phase shift can be retrieved as calculated from Eq. (1) when a rotation-rate Ω is applied to the whole system. Also, as showed in **Figure 3**, other elements like a second beam splitter, a polarizer and an optical filter are needed to complete the system. The main function of the electro-optical phase modulator is to provide a controlled phase shift which will be added to Sagnac phase shift produced by the rotation onto the system. This way, the signal detected by photo-detector can be demodulated with some ease to recover by electronic means the Ω rotation-rate value which affects the whole system.

A more advanced design is achieved by closing the measurement loop by means of a feedback signal becoming into the scheme so-called IFOG closed loop configuration. The general scheme of a closed loop IFOG is depicted on **Figure 4**. In this scheme, the output signal of demodulator circuit passes through a servo amplifier

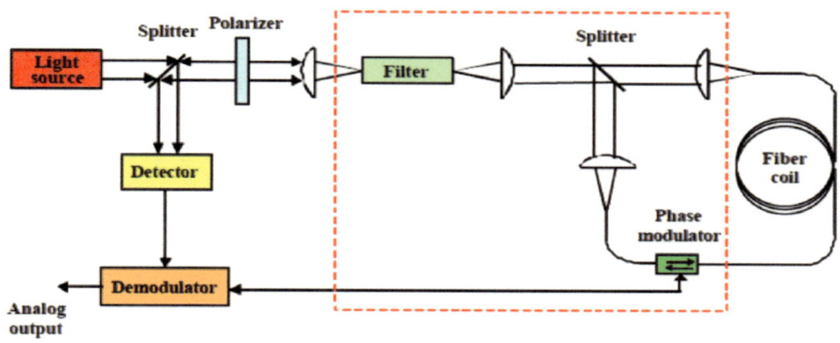

Figure 3.
Complete bulk-optics fiber-optic gyro open-loop configuration.

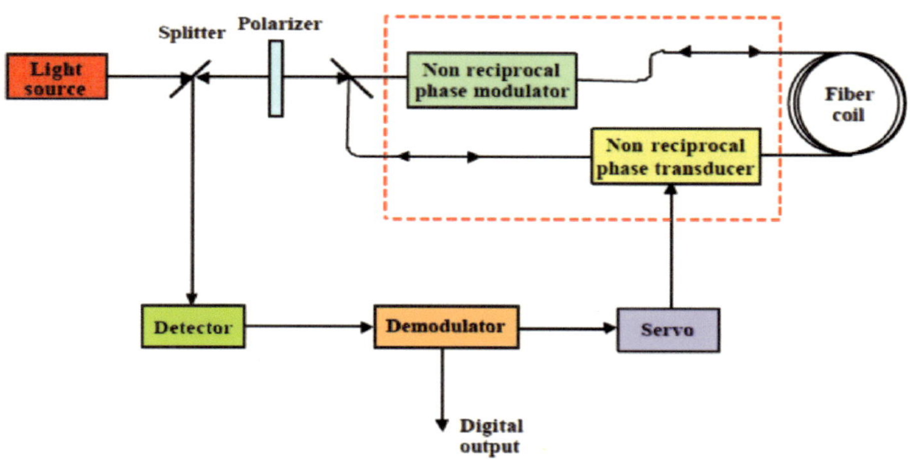

Figure 4.
Typical closed-loop IFOG configuration.

which drives a phase transducer placed in the interferometer path. Then, the whole system works under the phase-nulling principle. This means that the total phase shift becomes equal to zero because the phase transducer introduces a non-reciprocal phase shift that is equal, by in the opposite sign, to that generated by Sagnac phase shift induced by rotation. The output of the system is then the output of the phase transducer.

The main advantage of this configuration is the insensitivity to the laser source amplitude variations and the electronic circuitry gain because the system is always operated at zero total phase shift. Other design alternatives are possible, and so instead of using a fiber optic coil as a sensor it can be used a ring resonator integrated in a silicon waveguide, Ref. [17].

For open-loop configuration square-wave bias and sinusoidal phase modulation are usually applied while for closed-loop configuration, sinusoidal or square-wave bias and serrodyne feedback phase modulations are frequently used. In the following, a particular study of all these types of optical phase modulation will be made.

2. Square-wave optical phase modulation

One of the first attempts to apply the principle of phase modulation to CW-CCW optical waves in an optical gyroscope can be seen in Ref. [18]. In this case, a sinusoidal-wave phase modulation is applied mainly due first to the ease of finding fast bulk phase modulators in lithium niobate ($LiNbO_3$), Ref. [19], and also reliable electronic sine wave oscillators. However, square-wave is frequently used as bias phase modulation because it allows periodically shift the working point of the gyro to each one of $\pm\pi/2$ constant values, respectively. This last is due that when the central working point of the gyro is either $+\pi/2$ or $-\pi/2$, its sensitivity reach a maximum value, as it can be seen on **Figure 5**. Then, when the system is not subjected to rotation ($\Omega = 0$), the output response is a pectinate-shaped curve with a constant value. But, when the system is subjected to a non-zero rotation rate, the interferometric response output curve is also a Square-Wave whose peak-to-peak amplitude is proportional to the value of the rotation speed. The latter can also be also checked by observing in detail the **Figure 6**. In this figure, τ is the transit time of the CW and CCW optical waves over the fiber-coil length and ϕ_S is the Sagnac phase shift caused between them by rotation.

For this purpose, one phase modulator is located at the end of fiber coil, as represented in the scheme of **Figure 4**. Thus, the calculation of effective phase shift induced by the phase modulation process between CW and CCW optical waves at the output of fiber coil after their respective roundtrip can be expressed as follows:

$$\Delta\phi(t) = \phi_{CCW}(t) - \phi_{CW}(t) = \phi(t) - \phi(t - \tau) \quad (3)$$

In Eq. (3), $\phi(t)$ represents the time waveform of applied phase modulation and τ is, again, the transit time around the fiber coil which, in time, can be calculated as:

$$\tau = \frac{nL}{c} \quad (4)$$

here, n is the effective refractive index of fiber, L is the total length of fiber coil and c is the vacuum speed of light. For obtaining the result of Eq. (3) it has been taken into account that CW and CCW waves enter the fiber coil at opposite ends. Then, in the case of square-wave phase modulation as represented in **Figure 6**,

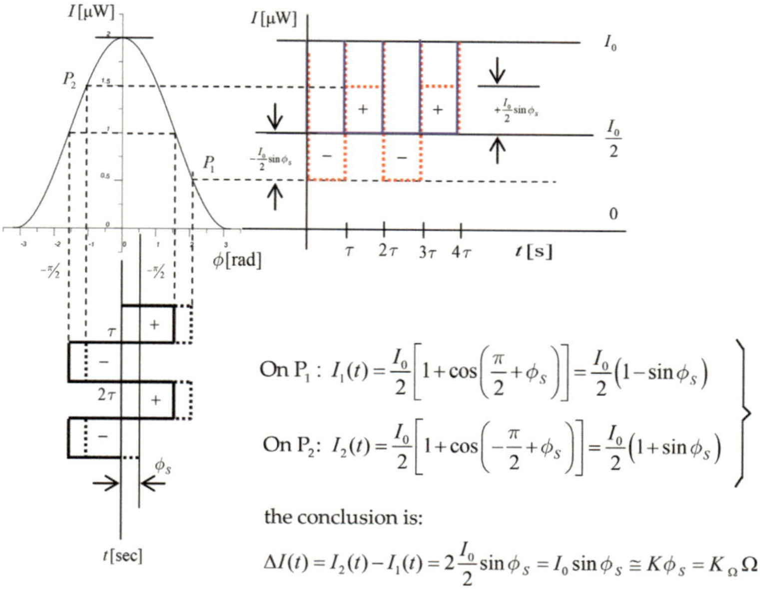

Figure 5.
Analysis of the interference response of the gyro with square-wave phase modulation (solid blue curve: zero rotation, dotted red curve: non-zero rotation).

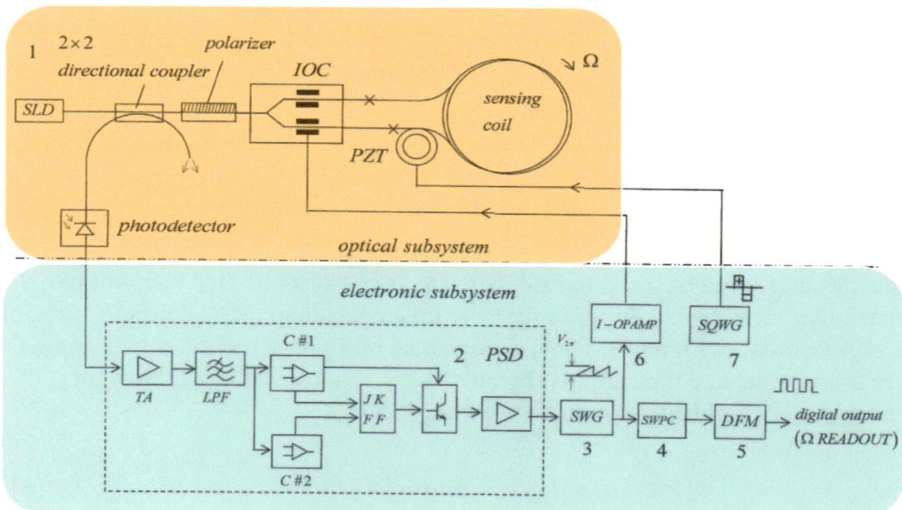

1- Optical Subsystem
2- Phase-Sensitive-Demodulator (PSD)
3- Sawtooth-Wave-Generator (SWG)
4- Sawtooth-Wave-to-Pulse-Converter (SWPC)
5- Digital-Frequency-Meter (DFM)
6- Inverter-Operational- Amplifier (I-OPAMP)
7- Square-Wave-Generator (SQWG)

Figure 6.
Closed-loop IFOG scheme with square-wave BIAS and serrodyne FEEDBACK phase modulations (below, in the inset, the block-diagram definition).

analytics gives two constant phase values, namely, $\pm\pi/2$ (continuous black wave) when no-rotation is applied to the system. However, when a non-zero rotation is applied, a phase difference equal to ϕ_S Sagnac phase shift must be added to phase difference applied by the external phase modulation process. The explanation of blue (continuous) and red (dotted) output response curves of photo-detector is as follows. When rotation rate is equal to zero, the projection of the points of the input square waveform (continuous black) on the response curve of the interferometer gets a pectinate-shaped output waveform (continuous blue curve). However, when a non-zero rotation rate, a pectinate square-wave is obtained (dotted red curve). In this last case, as seen in the inset of **Figure 6**, the peak-to-peak square-wave value is very close proportional to rotation rate and can be evaluated as:

$$\Delta I(t) = I_0 \sin \phi_S \cong K\phi_S = K_\Omega \Omega \tag{5}$$

where the approximation can be justified because the value of the sine-function can be approximated by its argument when it is less than π/6 in absolute value.

Figure 6 shows a closed-loop block-diagram scheme of an IFOG model with square-wave BIAS and serrodyne FEEDBACK phase modulations. In this case it is necessary to use two different phase modulators, first one to apply the square wave BIAS and second one for the serrodyne FEEDBACK phase modulation. See the complete description of this block-diagram on Ref. [20].

Gyro designs with square-wave kind of phase modulation can be seen in Refs. [21–24]. Last three engineered by Chinese authors utilize square waves staggered by sections (four-, five- or six-points phase modulation, respectively) and their main advantage is that all these schemes allow to improve the accuracy and scale factor of the gyro.

3. Sinusoidal optical phase modulation

The basic idea to apply the sinusoidal bias phase modulation to an IFOG configuration is that the amplitude of the first harmonic component of interferometer output signal contains information of the ϕ_S Sagnac phase shift induced by rotation. In particular, this amplitude can be considered approximately linearly proportional to the absolute value of rotation rate that affects the system. This fact will be analytically derived next. Suppose first a simple open-loop IFOG configuration, like that represented on **Figure 3**. Then, a sinusoidal bias phase modulation is applied to phase modulator, like the supplied by an electric sine oscillator working at f_m frequency and amplitude ϕ_0 in the following form:

$$\phi_m(t) = \phi_0 \sin(\omega_m t) \tag{6}$$

The phase difference between CW and CCW waves induced by this bias phase modulation will be:

$$\begin{aligned} \Delta\phi_m(t) &= \phi_{CCW}(t) - \phi_{CW}(t) = \phi_m(t) - \phi_m(t-\tau) \\ \Delta\phi_m(t) &= 2\phi_0 \sin\left(\frac{\omega_m \tau}{2}\right) \cos\left[\omega_m\left(t - \frac{\tau}{2}\right)\right] \\ &\xrightarrow{\omega_m \tau = \pi} 2\phi_0 \cos\left[\omega_m\left(t - \frac{\tau}{2}\right)\right] = \phi_m \sin(\omega_m t) \end{aligned} \tag{7}$$

here $2\phi_0 = \phi_m$. As it can be seen from this equation, the maximum value of phase difference modulation for a given value of ϕ_0 amplitude will be reached when the $\omega_m \tau = \pi$ condition to be accomplished. This condition is reached when the frequency f_m equals the value:

$$f_m = \frac{1}{2\tau} \tag{8}$$

referred as proper frequency of the system, here τ, the transit time. Then, under these conditions, when a rotation with ϕ_S Sagnac phase shift affects the system, the total phase difference between CW and CCW waves will be:

$$\Delta\phi(t) \;=\; \Delta\phi_m(t) + \phi_S = \phi_m \sin(\omega_m t) + \phi_S \tag{9}$$

Therefore, the interference signal can be obtained by the following calculation:

$$\begin{aligned} I(\Delta\phi) &= I_0[1 + \cos(\Delta\phi)] = I_0[1 + \cos[\phi_m \sin(\omega_m t) + \phi_S]] \\ &= I_0 \left\{ \begin{aligned} &1 + \left[J_0(\phi_m) + 2\sum_0^{\infty} J_{2n}(\phi_m)\cos(2n\,\omega_m t)\right]\cos(\phi_S) \\ &-2\sum_0^{\infty} J_{2n-1}(\phi_m)\sin[(2n-1)(\omega_m t)]\sin(\phi_S) \end{aligned} \right\} \end{aligned} \tag{10}$$

here J_n is the Bessel function of first kind and n-order. From Eq. (10) it can be observed that interference signal is the sum of three terms: a constant first term that does not depend on the frequency of modulation, a second term that includes the factor $\cos(\phi_S)$ multiplied by one infinite sum of even harmonics and a third term that includes the factor $\sin(\phi_S)$ multiplied by one infinite sum of odd harmonics. When ϕ_m value is adjusted to be 1.84, then the J_1 term reach its maximum value, namely, $J_1(1.84) = 0.5815$. It has also been shown that the amplitude of higher-order odd harmonics (3rd, 5th and successive) are getting smaller and smaller so their contribution can be neglected. Therefore, when a rotation rate affects the system, an electronic selective filtering of odd harmonics of interference signal is convenient to isolate and extract the data of rotation speed. Particularly, the first harmonic is the most convenient to recover, so that an electronic band-pass filtering at fundamental frequency of modulation should be added as phase-sensitive demodulator. This way, a successive low-pass filtering located after the band-pass one could retrieve the first harmonic amplitude, which is proportional to ϕ_S Sagnac phase shift and, in turn, to rotation rate value.

Figure 7 represents the analytic sinusoidal phase modulation process. As it can be seen, when no rotation is applied, the interference signal (solid, blue curve) also contains even harmonics since the $\sin(\phi_S)$ factor is canceled. However, when rotation is applied, both even and odd harmonics appear (dotted, red curve) in the interference output signal since the $\sin(\phi_S)$ and $\cos(\phi_S)$ are not canceled.

Figure 8 represents the block-diagram of a closed-loop IFOG configuration with sinusoidal BIAS and serrodyne FEED-BACK phase modulations, see Ref. [25]. The main novelty of this design is the structure of phase modulation feedback chain. In this case, one FET transistor (2N3848) is added on feedback branch of integrator OPAMP (block #7 on **Figure 8**). This block generates a linear ramp voltage $V\gamma$ on its output, and this ramp resets each one time-period driving by V_{gate} voltage. In this way, a resultant serrodyne-wave voltage is easily generated at the output of integrator circuit, obtaining finally the same intended sawtooth-wave voltage on feedback phase modulation chain as reported on previous designs. The output signal of the photodetector, in photocurrent form, is proportional to the light intensity at its optical input. This photocurrent signal is converted to voltage with a trans-impedance amplifier that is placed at the input of demodulation circuit. The demodulation circuit (PSD) takes the task of extracting the information of the ϕ_S Sagnac phase shift induced by rotation. The corresponding voltage signal at its

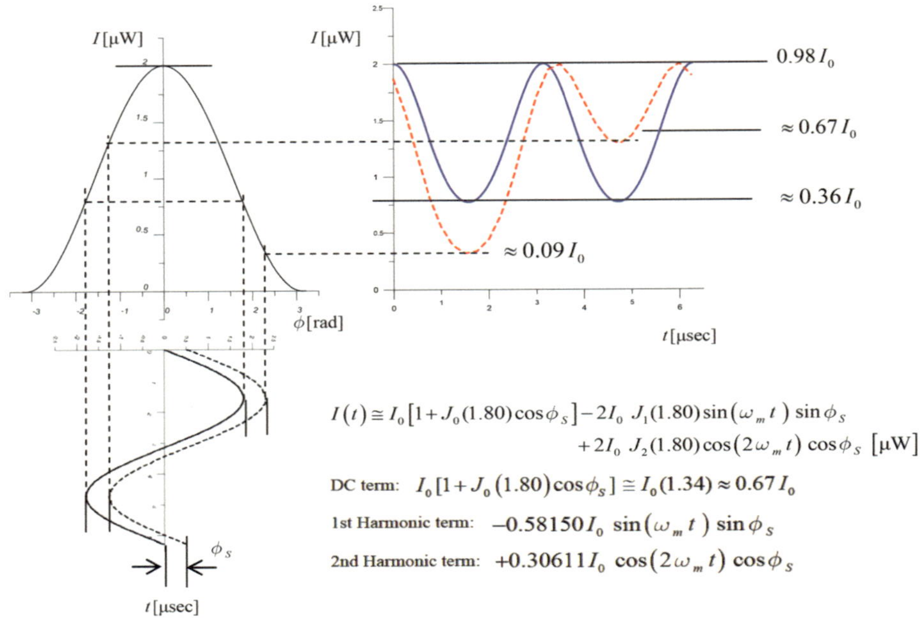

Figure 7.
Analysis of the interference response of the gyro with sinusoidal phase modulation (solid blue curve: zero rotation, dotted red curve: non-zero rotation).

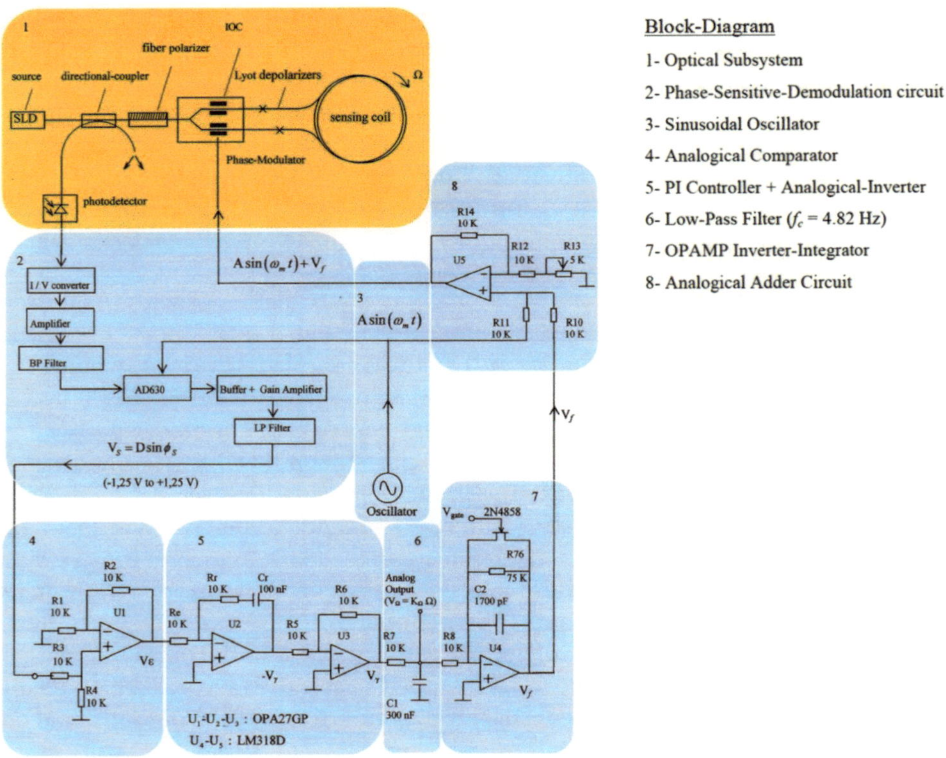

Figure 8.
Closed-loop IFOG scheme with sinusoidal BIAS and serrodyne FEEDBACK phase modulations, see Ref. [25] (below, in the inset, the block-diagram definition).

output (V_S) scales as sine-function of ϕ_S. The PI controller realizes an integration of V_S signal in time-domain, so that a voltage signal (Vγ) is obtained; this signal scales almost linearly with the time. This latter signal is filtered by means of a low-pass filter so that the corresponding output signal (V_Ω) is a DC voltage value that is more accurately proportional to the gyroscope rotation-rate Ω (since the following approximation is fulfilled in the working range: $\sin\phi_S \approx \phi_S$). Therefore, the V_Ω analog output voltage signal constitutes the measurement of the rotation rate of the system. The control system, as a whole, acts as the principle of phase nulling. The phase-nulling process consists of generating a phase displacement ($\phi_m = \phi_{bias} + \phi_f$) in such a way that the ϕ_f phase-difference associated with the voltage output signal (V_f) is equal and with opposite sign with regard to the Sagnac phase-shift induced by the rotation rate, i.e., $\phi_f = -\phi_S$. To achieve this, the feedback phase modulation circuit holds a sample of the output signal V_Ω. Note that this voltage signal is obtained at the end of low pass filter (Block #6 on **Figure 8**) and is proportional to rotation-rate Ω. An integration operation is needed for obtaining a linear ramp voltage to apply on phase modulator. Then, it integrates and inverts this signal by means of an operational integrator-inverter circuit, turning this signal into the following form:

$$V_f = -\frac{1}{RC}\int_0^t V_\Omega dt \tag{11}$$

This way, the time variation of V_f voltage signal is a linear ramp, being its slope proportional to the rotation rate of the system (V_Ω). **Figure 8** represents clearly the optical and electronic subsystems of the gyroscope, including the feedback phase-modulation and bias phase-modulation circuits for getting phase-nulling process, both applied together to PM (phase-modulator). Referring now to **Figure 8**, the total voltage signal applied to PM will be:

$$V_m = V_{bias} + V_f \tag{12}$$

Therefore, the output signal of the phase modulator will be the sum of the phase-difference signals associated with the V_{bias} and V_f voltages. In terms of phase differences, this is expressed as $\phi_m = \phi_{bias} + \phi_f$. Then, the error signal at the output of the comparator ($V\varepsilon$ voltage) tends to be nulled in average-time, due to the phase cancelation (since the average-time of the reference bias phase-modulation ϕ_{bias} is 0).

Sinusoidal phase modulation has been used on either open-loop (Refs. [18, 26, 27]) or closed-loop (Refs. [25, 28–30]) IFOG configurations. In both cases, the PSD block (phase-sensitive-demodulation) must contain two selectively adjusted filtering circuits. First one is a low-pass filter with high enough cut-off frequency to filter the first harmonic component. Second one is a selective band-pass-filter to filter the component of first harmonic of interference signal, see also Ref. [25].

4. Serrodyne optical phase modulation

As announced at the beginning of this work, serrodyne-wave optical phase modulation is frequently used in closed loop IFOG schemes to configure one feedback signal which is able to cancel the ϕ_S Sagnac phase shift induced by rotation. The justification for this is that the serrodyne-wave is the only one that produces a constant phase difference when applied to phase modulator (PM) in a gyro. This last

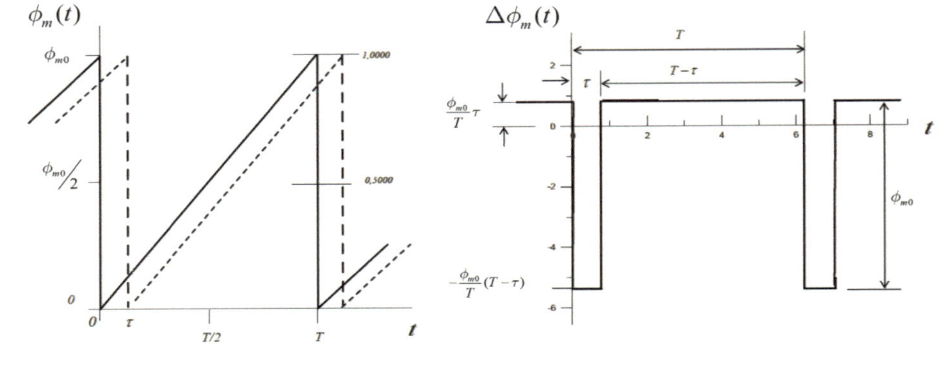

(a) (b)

Figure 9.
Serrodyne-wave phase modulation: (a) serrodyne-wave applied to PM with amplitude ϕ_{m0} and period T and (b) effective phase-difference $\Delta\phi_m$(t)as a function of time.

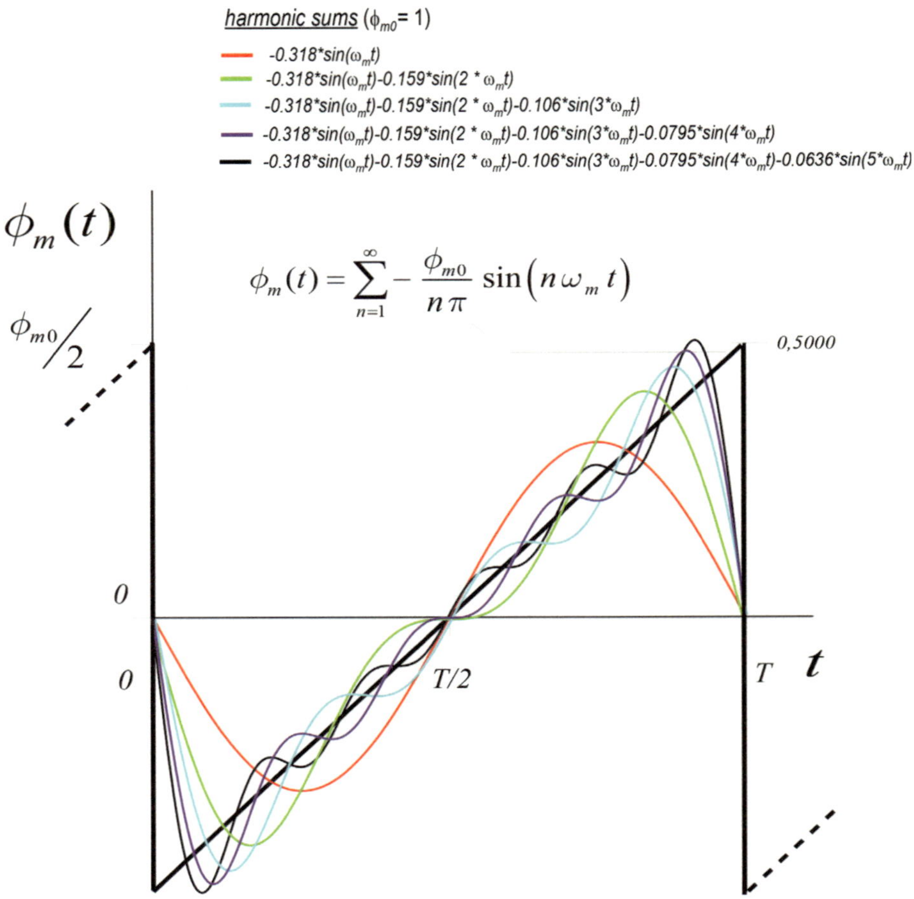

Figure 10.
Fourier series development (Ref. [34]) of the sawtooth wave. In the upper inset, sum series contain the following: 1st harmonic (red curve); 1st and 2nd harmonics (blue curve); 1st, 2nd, and 3rd harmonics (green curve); 1st, 2nd, 3rd, and 4th harmonics (cyan curve); and finally, 1st, 2nd, 3rd, 4th, and 5th harmonics (black curve).

can be checked observing the **Figure 9(b)**. During a time span equal to $(T - \tau)$ the phase-difference between CW and CCW waves remains constant with a value equal to $\frac{\phi_{m0}}{T}\tau$, being τ the transit time of light around the fiber coil. In addition, a constant value of 2π is usually taken as the ϕ_{m0} amplitude of phase modulation in most part of designs. This way, by adjusting appropriately the period T (or the frequency) of serrodyne-wave, the resulting value of phase-difference can be exactly matched with the ϕ_S Sagnac phase shift to achieve the phase cancelation by means of a specific feed-back circuit located on the way of feed-back signal, see for example Refs. [25, 31–33].

For a proper operation of feedback circuit, it is essential that the falling edge (reset time) of sawtooth-wave be as fast as possible (ideally instantaneous), Ref. [31]. Since that serrodyne- (or sawtooth-) referred waveform is a periodic waveform that accomplishes Dirichlet conditions in the $(0,T)$ interval, it is susceptible to be developed in a Fourier series (Ref. [34]) such as the one following next:

$$\phi_m(t) = \sum_{n=1}^{\infty} -\frac{\phi_{m0}}{n\pi}\sin\left(n\omega_m t\right) \tag{13}$$

This result shows that the series only contains sine terms because it refers to an odd function. On the other hand that result is very useful for filtering design purposes as it can be seen on simulated plots represented in **Figure 10**. Here, a successive sums containing the harmonics: first one (red-curve), the first and second ones (green-curve), the first, second and third ones (blue-curve), the first, second, third and fourth ones (cyan-curve) and finally the first, second, third, fourth and fifth ones (black-curve) are represented. The more terms are taken from the sum series, the better the approximation will be to the perfect sawtooth waveform.

In order to realize the serrodyne-wave phase modulation an Voltage-Controlled-Oscillator (VCO) circuit must be designed. The condition that this circuit must comply is:

$$\phi_S = \frac{V_{2\pi}}{T}\tau \tag{14}$$

so that the frequency $f = 1/T$ of serrodyne-wave should be adjusted depending on the value of the ϕ_S Sagnac phase shift, i.e., the more be ϕ_S, the more will be the frequency of serrodyne and, then, the lower the value of its period T. Several circuits have been designed to meet this condition. One of these circuits has been represented in **Figure 11** and is described in Ref. [35]. Other VCO circuit for serrodyne-wave generation has already been explained above for FEED-BACK phase modulation, see **Figure 8** and related Ref. [25].

Although the serrodyne wave is the one that produces the best results for the feed-back phase modulation purpose, other similar waves have been also proposed. For example, symmetric triangular-wave represented in **Figure 12** can also perform the same function. Since it is an odd function, its development in Fourier series only contains the odd harmonics, then, it can be expressed in the following way:

$$\phi_m(t) = 4\,\phi_{m0}\sum_{n=1}^{\infty}\frac{\sin\left(\frac{n\pi}{2}\right)}{n^2\pi^2}\sin\left(\frac{n\pi t}{T/2}\right) \quad n = 1,3,5,\dots \tag{15}$$

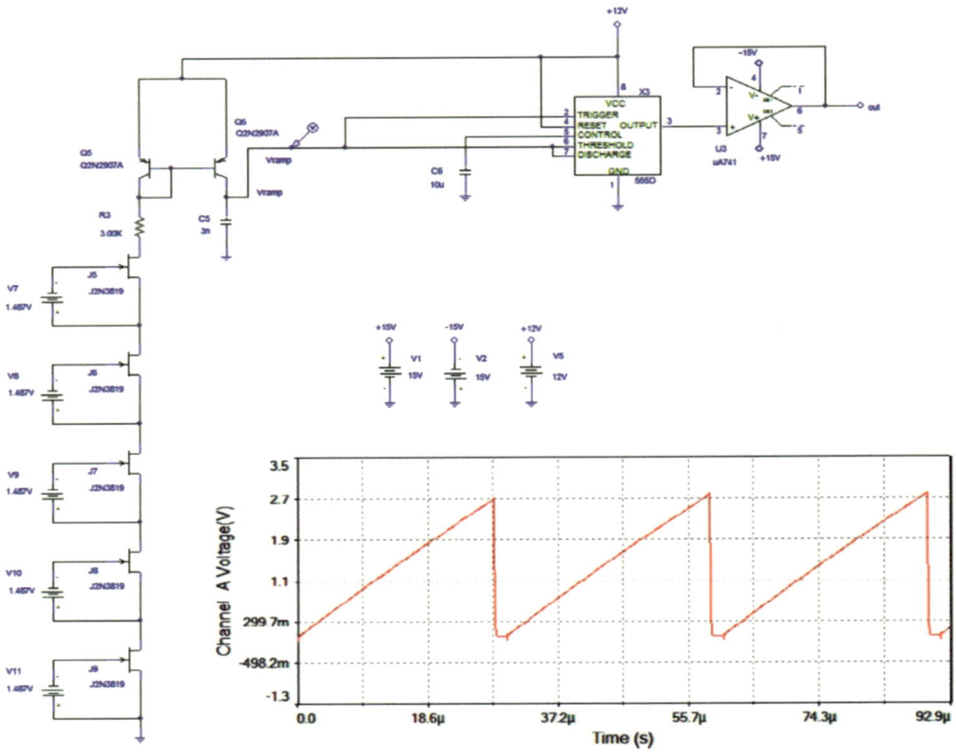

Figure 11.
Serrodyne-wave VCO (voltage-controlled-oscillator) circuit with 555 IC (in the inset, 35.84 kHz linear ramp waveform generated), Ref. [35].

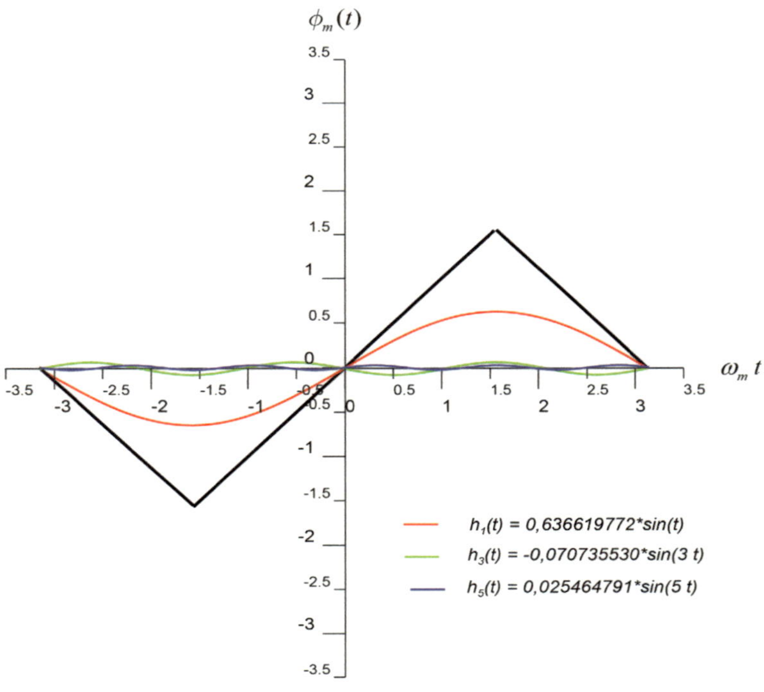

Figure 12.
Symmetric triangular-wave with its 1st, 2nd and 3th harmonics, Ref. [35].

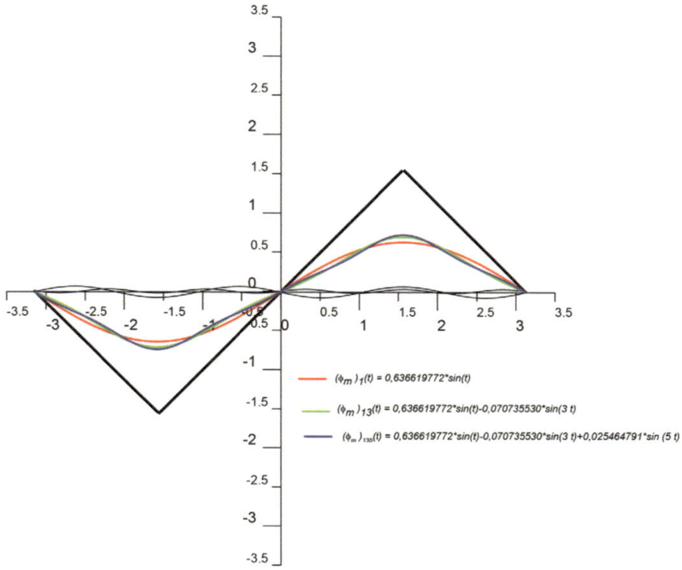

Figure 13.
Symmetric triangular wave (in black) with its first three harmonic Fourier sums (red, green, blue), Ref. [35].

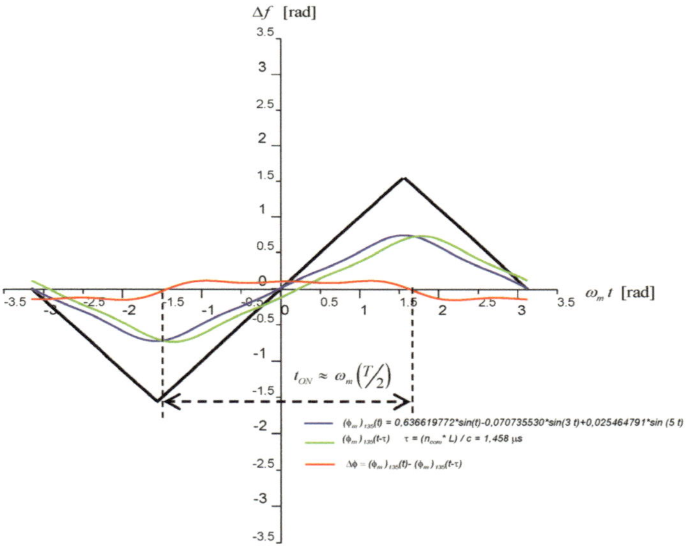

Figure 14.
Effective phase-difference modulation of symmetric triangular wave (red-curve) related with its first three harmonic Fourier sums (blue, green) of CCW and CW optical waves, Ref [35].

Figure 13 represents the three first harmonic sums (red, green, blue) of symmetric triangular-wave (black). So that taking $\phi_{m0} = \pi/2$ in Eq. (15), the first three harmonic terms can be written as follows:

$$h_1(t) = \frac{4\phi_{m0}}{\pi^2} \sin\left(\frac{1\,\pi}{2}\right) \sin\left(\frac{\pi\, t}{T/2}\right) = 0{,}636619772 \sin(t) \tag{16}$$

$$h_3(t) = \frac{4\phi_{m0}}{9\pi^2} \sin\left(\frac{3\pi}{2}\right) \sin\left(\frac{3\pi\, t}{T/2}\right) = -0{,}070735530 \sin(3\,t) \tag{17}$$

$$h_5(t) = \frac{4\phi_{m0}}{9\pi^2} \sin\left(\frac{5\pi}{2}\right) \sin\left(\frac{5\pi t}{T/2}\right) = 0,025464791 \sin(5\ t) \tag{18}$$

This way, the first three harmonic Fourier sums can be expressed, respectively, as:

$$(\phi_m)_1 (t) = \frac{4\phi_{m0}}{\pi^2} \sin\left(\frac{\pi t}{T/2}\right) \tag{19}$$

$$(\phi_m)_{13} (t) = \frac{4\phi_{m0}}{\pi^2} \left[\sin\left(\frac{\pi t}{T/2}\right) - \frac{1}{9} \sin\left(\frac{3\pi t}{T/2}\right) \right] \tag{20}$$

$$(\phi_m)_{135} (t) = \frac{4\phi_{m0}}{\pi^2} \left[\sin\left(\frac{\pi t}{T/2}\right) - \frac{1}{9} \sin\left(\frac{3\pi t}{T/2}\right) + \frac{1}{25} \sin\left(\frac{3\pi t}{T/2}\right) \right] \tag{21}$$

Then, the conclusion is that when the approximation of first three harmonic Fourier sum of symmetrical triangular-wave is taken in gyro phase modulation (blue and green curves in **Figure 14**), its effective phase-difference (red curve in **Figure 14**) can be computed. In this case, one switching circuit is needed.

5. Conclusions

Square, sinusoidal, serrodyne and symmetric triangular waveforms can be used in phase modulation processes for optical Sagnac interferometer gyros. For open-loop gyro schemes only one square-wave or sinusoid$$PM) to retrieve the Sagnac phase shift induced by rotation. However, for closed-loop scheme gyros two waveforms are needed, the first one (square-wave or sinusoidal-wave) for the bias phase modulation and the second one (serrodyne-wave or triangular-wave) for the feedback phase modulation aiming the phase cancellation (phase-nulling). In the closed-loop scheme, the output signal of the phase-sensitive-demodulator (PSD) circuit passes through a servo-amplifier which drives a phase-shifter transducer placed in the interferometer path. Then, the phase transducer introduces a non-reciprocal phase shift that is equal, by in the opposite sign, to Sagnac phase shift induced by rotation. Thus, the output of the system is the output of the phase transducer. Closed-loop gyro configuration is advantageous with regard the open-loop one because a better accuracy (sensitivity) and scale-factor stability of the gyro are achieved.

Author details

Ramón José Pérez Menéndez
UNED-Spain, Lugo, Spain

*Address all correspondence to: ramonjose.perez@lugo.uned.es

References

[1] Hariharan P. Optical Interferometry. 2nd ed. Academic Press; 2003. pp. 1-8

[2] Sharma U, Wei X. Fiber optic interferometric devices. In: Kang JU, editor. Fiber Optic Sensing and Imaging. New York: Springer; 2013. pp. 29-53 [chapter 2]

[3] Yin S, Ruffin PB, Yu FTS, editors. Fiber Optic Sensors. 2nd ed. CRC Press; 2008. pp. 333-366

[4] Vali V, Shorthill RW. Fiber ring interferometer. Applied Optics. 1976; **15**(5):1099-2000

[5] Bohm K, Marten P, Petermann K, Weidel E, Ulrich R. Low-drift using a superluminescent diode. Electronics Letters. 1981;**17**:352-353

[6] Ruffin PB, Smith RH. Fiber winding approaches for environmentally robust IFOG sensor coils. SPIE of Proceedings, 1792. Components for Fiber Optic Applications. 1992. p. 179

[7] Bergh RA, Lefevre HC, Shaw HJ. All-single-mode fiber-optic gyroscope. Optics Letters. 1981;**6**(4):198-200

[8] Bergh RA, Lefevre HC, Shaw HJ. Compensation of the optical Kerr effect in fiber-optic gyroscopes. Optics Letters. 1982;7:282-284

[9] Bergh RA, Lefevre HC, Shaw HJ. Overview of fiber-optic gyroscopes. IEEE Journal of Lightwave Technology. 1984;**LT-2**:91-107

[10] Moeller RP, Burns WK, Frigo NJ. Open-loop output and scale factor stability in a fiber-optic gyroscope. Journal of Lightwave Technology. 1989; 7(2):262-269

[11] Hotate K, Tabe K. Drift of an optical fiber gyroscope caused by the faraday effect: Influence of the earth's magnetic field. Applied Optics. 1986;**25**: 1086-1092

[12] Bohm K, Petermann K, Weidel E. Sensitivity of a fiber optic gyroscope to environmental magnetic fields. Optics Letters. 1982;**6**:180-182

[13] Bergh RA, Culshaw B, Cutler CC, Lefevre HC, Shaw HJ. Source statistics and the Kerr effect in fiber-optic gyroscopes. Optics Letters. 1982;**7**: 563-565

[14] Takiguchi K, Hotate K. Method to reduce the optical Kerr-effec-induced bias in an optical passive ring-resonator gyro. IEEE Photonics Technology Letters. 1992;**4**:2

[15] Shupe DM. Thermally induced nonreciprocity in the fiber-optic interferometer. Applied Optics. 1980; **19**(5):654-655

[16] Ruffin PB, Lofts CM, Sung CC, Page JL. Reduction of nonreciprocity in wound fiber optic interferometers. Optical Engineering. 1994;**33**(8): 2675-2679

[17] Lefevre HC, Bettini JP, Vatoux S, Papuchon M. Progress in optical fiber gyroscopes using integrated optics. In: NATO-AGARD Conference Proceedings, 383, 9A1, 9A-13. 1985

[18] Kim BY, Shaw HJ. Phase-reading all-fiber-optic gyroscope. Optics Letters. 1984;**9**(8):378-380

[19] Wooten L, Kissa KM, Yi-Yan A, Murphy EJ, Lafaw DA, Hallemeier PF, et al. A review of lithium niobate modulators for fiber-optic communications systems. IEEE Journal of Quantum Electronics. 2000;**6**(1):69-81

[20] Pérez Menéndez RJ. Optoelectronic design of a 2045 m coil, closed loop-

based, depolarized IFOG with square-wave bias and Sawtooth-wave feedback optical phase modulations: Parametric modelling, simulation and performance test. In: Yurish SY, editor. Book Series: Advances in Sensors: Reviews. Vol. 5. Barcelona, Spain: IFSA Publishing S.L.; 2018

[21] Kay CJ. Serrodyne modulator in a fibre-optic gyroscope. IEE Proceedings Journal—Optoelectronics. 1985;**132**(5): 259-264

[22] Li X, Zhang Y, Yu Q. Four-state modulation in fiber optic gyro. In: 2008 IEEE International Conference on Mechatronics and Automation; Takamatsu. 2008. pp. 189-192

[23] Li X, Zhang Y, Zhang C. Five points modulation in closed loop fiber optic gyroscope. In: 2009 5th International Conference on Wireless Communications, Networking and Mobile Computing. Beijing; 2009. pp. 1-3

[24] Zhang C, Zhang S, Pan X, Jin J. Six-state phase modulation for reduced crosstalk in a fiber optic gyroscope. Optics Express. 2018;**26**:10535-10549

[25] Pérez Menéndez RJ. Optoelectronic design of a closed-loop depolarized IFOG with sinusoidal phase modulation for intermediate grade applications. 2018. pp. 215-223. DOI: 10.5772/intechopen.72592

[26] Emge SR, Bennett S, Dyott RB, Brunner J, Allen DE. Reduced minimum configuration fiber optic gyro for land navigation applications. Aerospace and Electronic Systems Magazine. IEEE; 1997;**16**:18-21

[27] Bennett SM, Emge S, Dyott RB. Fiber optic gyroscopes for vehicular use. In: Proceedings of Conference on Intelligent Transportation Systems. Boston, MA, USA; 1997. pp. 1053-1057

[28] Kim BY, Shaw HJ. Harmonic feed-back approach to fiber gyroscope scale factor stabilization. In: Proceedings. 1st Conference. Optical Fiber Sensors. London; 1983. pp. 136-137

[29] Kim BY, Shaw HJ. Gated phase-modulation feedback approach to fiber-optic gyroscopes. Optics Lettters. 1984;**9**(6):263-265

[30] Kim BY, Shaw HJ. Gated phase-modulation feedback approach to fiber-optic gyroscopes with linearized scale factor. Optics Lettters. 1984;**9**(8): 375-377

[31] Kay CJ et al. Serrodyne modulator in a fibre-optic gyroscope. IEEE Proceedings Journal. 1985;**132**(5):259-264

[32] Ebberg A et al. Closed-loop fiber-optic gyroscope with a Sawtooth phase-modulated feedback. Optics Letters. 1985;**10**(6):300-302

[33] Skalský M, Habránek Z, Fialka J. Efficient modulation and processing method for closed-loop fiber optic gyroscope with piezoelectric modulator. Sensors (MDPI). 2019;**19**(7):1710

[34] Abramovitz M, Stegun I. Handbook of Mathematical Functions: With Formulas, Graphs and Mathematical Tables. Dover Publications; 1972. p 805

[35] Pérez RJ. Development of prototypes of single-mode fiber-optic interferometric sensor with phase modulation techniques for the measurement of differential rotation [PhD thesis]. Universidad de Oviedo-Spain. 2016. pp. 299-305. Available from: http://digibuo.uniovi.es/dspace/handle/10651/40188

Chapter 6

Review of Injected Oscillators

Ali Reza Hazeri

Abstract

Oscillators are critical components in electrical and electronic engineering and other engineering and sciences. Oscillators are classified as free-running oscillators and injected oscillators. This chapter describes the background necessary for the analysis and design of injected oscillators. When an oscillator is injected by an external periodic signal mentioned as an injection signal, it is called an injected oscillator. Consequently, two phenomena occur in the injected oscillators: (I) pulling phenomena and (II) locking phenomena. For locking phenomena, the oscillation frequency of the injection signal must be near free-running oscillation frequency or its sub-/super-harmonics. Due to these phenomena are nonlinear phenomena, it is tough to achieve the exact equation or closed-form equation of them. Therefore, researchers are scrutinizing them by different analytical and numerical methods for accomplishing an exact inside view of their performances. In this chapter, injected oscillators are investigated in two main subjects: first, analytical methods on locking and pulling phenomena are reviewed, and second, applications of injected oscillators are reviewed such as injection-locked frequency dividers at the latter. Furthermore, methods of enhancing the locking range are introduced.

Keywords: frequency dividers, frequency multipliers, injected oscillators, nonlinear oscillations, locking phenomena, locking range, quadrature oscillators, pulling phenomena

1. Introduction

One of the most important blocks in the electronic systems is undoubtedly the oscillator block. The oscillator converts a direct current (DC) generated by power supply into an alternating current (AC) signal. The design of this block involves many trade-offs between phase noise, oscillation frequency range, power consumption, layout size, etc. Oscillators are employed in many applications such as phase-locked loops (PLL), frequency dividers/multipliers, clock recovery, frequency synthesizers, etc. Nowadays, the demand for compact and portable systems has increased. Hence, the oscillators, amplifiers, mixers, power amplifiers, etc. may be integrated on a chip. For example, FM radio, Bluetooth, GPRS, Wi-Fi, NFC, and GPS are integrated into modern mobile systems. According to the application, the oscillators are classified as the free-running oscillators and injected oscillators. Free-running oscillators have been extremely studied in many electronic engineering books and articles so far [1–6]. In the free-running oscillators, there is no external signal injected into the oscillator. In the injected oscillators, an external periodic signal mentioned as an injection signal is injected to the oscillator, which may be deliberately applied by the designers to make an injected oscillator, or any injection signals are accidentally injected to the oscillator which may be generated from other blocks such as power amplifiers and other oscillators. By injected oscillators, designers can implement many high-performance blocks such as quadrature oscillators and frequency dividers/multipliers, frequency synthesizers without a frequency-locked loop, which are useful in the fast frequency locking-loop systems. Thus, the cost of fabrication is cheaper than a frequency-locked loop. There are many devices operating with different center frequencies. As a result, several oscillators and other devices are placed together for implementing a system called system on chip (SOC). However, when the oscillators are integrated with other devices on the chip, various signals with different center frequencies may leak through the substrate, parasitic elements, or packaging and be injected into the oscillator. Hence, the performance of the oscillators is changed which can be suitable or not dependent on their functionality. When an oscillator is injected with an injection signal, the pulling and locking phenomena occur. Thus, pulling and locking phenomena are important parameters for designers.

The injected locked oscillators designed by engineers are classified into three classes. While the oscillation frequency of the injection signal is near to the free-running oscillator, the first-harmonic injection locking takes place. When the oscillation frequency of the injection signal is near to the sub-/super-harmonic of the oscillation frequency of the free-running oscillator, frequency dividers/multipliers are realized. These phenomena occur since the nonlinear performance of the injected oscillators. So, the size of the layout and complexity of designing frequency dividers/multipliers are lower than the frequency-locked loop because they do not employ many blocks in the frequency-locked loop such as filters, charge pomp, and frequency detector [7]. Consequently, power consumption is reduced. Furthermore, they are fast and may be applied to the high-speed or high clock data recovery and fast-locking systems. At last, the phase noise of the injected oscillator is different from the free-running oscillator and is dependent on the phase noise of the injection signal. Therefore, while the injection signal is generated by an oscillator which owns excellent phase noise, the injected oscillator will have a better phase noise [7, 8]. This chapter tries to disclose all subtleties and challenges encountered during the design of injected oscillators.

The presented chapter aims to investigate the injected oscillators. A summary of the injected oscillator specifications regarding locking and pulling phenomena, previously significantly published papers about the first harmonic injection oscillator, frequency dividers and enhancing locking range are presented. Section 2 covers with introducing pulling and locking phenomena in the first-harmonic injection locking oscillator. Then, a free-running oscillator is implemented for exploring the pulling and locking phenomena by various injection signals. Section 3 will be dedicated to the pulling and locking formula, and beat frequency equation is discussed for injected oscillators for nonharmonic (LC) and harmonic (ring) oscillators. Section 4 will treat the implementation of injected locked frequency dividers and increase the locking range. First, a block of the injection-locked frequency dividers/multiplier is displayed. In a literature overview, two classes of realization will be recognized: the conventional LC-injection-locked frequency dividers where the injection signal is applied to the oscillator by the tail current source or a transistor connected to the output nodes. Furthermore, the previously important published papers are reviewed. Finally, some structures and techniques employed in order to extend the locking range of frequency dividers are exhibited from previously significant published papers. Section 5 will conclude with the main contributions of the presented chapter.

2. Pulling and clocking phenomena

When an oscillator is injected with an external signal, two phenomena occur for the oscillator. Once the strength of the external signal and frequency difference between the external signal and free-running oscillator are not suitable for locking, the oscillator is perturbed, and the output signal is modulated called pulling phenomenon. In the pulling case, the output spectrum of the injected oscillators has some spurious tones along with the effective oscillation frequency of the external signal. While the strength of the external signal and frequency difference between the external signal and free-running oscillator are suitable, the oscillator is locked, and the frequency of the output signal is locked to the first, sub- or super-harmonic of the oscillation frequency of the external signal called locking phenomenon. The range of the oscillation frequency of the external signal causes a locking phenomenon called the locking range. In **Figures 1** and **2**, these phenomena are simply presented in the time domain and frequency domain.

At first, a free-running oscillator is chosen. Next, the oscillation frequency, the output voltage oscillation amplitude, and the current oscillation amplitude are obtained. Second, the given oscillator is injected with the external signal, which is modeled by a current source and in parallel with the output nodes. Third, the results of the injected oscillator are revealed under locking and pulling phenomena.

The free-running oscillator, implemented by a three-stage differential ring oscillator, is depicted in **Figure 1a** along with the delay cell in **Figure 1b**. In the time and frequency domain, the output voltage of the oscillator is presented in **Figure 1c** and **d**. According to **Figure 1b** and **c**, there is no frequency or phase modulation on the output voltage. It has a center frequency (257 MHz). In addition, the output voltage amplitude is constant (0.481 V), and the oscillation current (I_{osc}) is equal to 0.18 mA. In fact, there is not any amplitude modulation (AM).

Figure 2a shows the given three-stage differential ring oscillator under an injection signal, which its current (I_{inj}) and oscillation frequency (F_{inj}) are equal to $I_{osc}/20$ and 260 MHz, respectively. Assuming locking conditions are covered, the oscillation frequency of the injected oscillator is the same as the injected signal as shown in **Figure 2b** and **c**. **Figure 2b** and **c** displays the output voltage of the injected oscillator under locking phenomenon in the time and frequency domain, respectively. **Figure 2b** and **c** illustrates the output voltage of the injected oscillator locked in F_{inj}. It is clear that there is not any modulation on the output voltage at the time and frequency domain. Then, the oscillation frequency of the injected oscillator is equal to the oscillation frequency of the injected signal.

Figure 2d–g illustrates the output voltage of the injected oscillator in the pulling phenomenon when $I_{inj} = I_{osc}/20$ and F_{inj} is equal to 262.5 and 265 MHz. It is obvious that there are modulations on the output voltage at the time and frequency domain. The injected oscillator produces amplitude and frequency (or phase) modulation in the output voltage. According to **Figure 2d** and **f**, the center frequency of the oscillator is pulled to the frequency of the external signal. Furthermore, the beat frequency, instance variation of oscillation frequency, is reduced when F_{inj} becomes near locking range as portrayed in **Figure 2**.

3. Review of the previously significantly published papers

The pulling and locking phenomena have been investigated in previously published papers [2, 8–60]. According to the output waveform, oscillators may be categorized as nonharmonic oscillators, for example, LC oscillators and harmonic oscillators such as ring oscillators. For the nonharmonic oscillators, the output waveform has a center frequency near the resonance frequency of the LC tank. Consequently, the output waveform is almost sinusoidal. For harmonic oscillators, since the output waveform is not sinusoidal, the higher harmonics effect on the output waveform. In fact, nonharmonic oscillators have an LC tank block

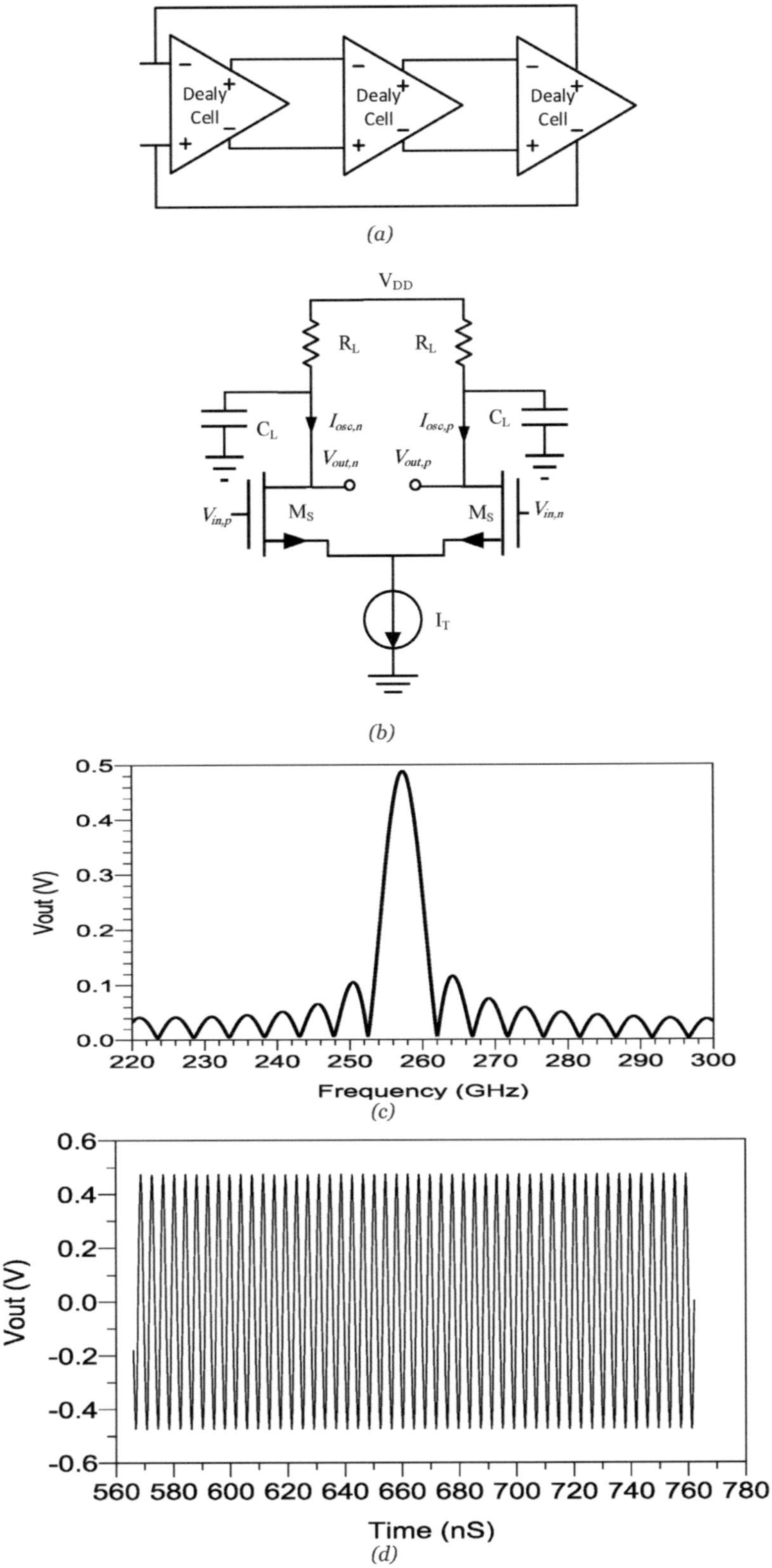

Figure 1.
(a) Three-stage differential ring oscillator, (b) delay cell, (c) output voltage at the time domain, (d) output voltage at the frequency domain.

DOI: http://dx.doi.org/10.5772/intechopen.91687

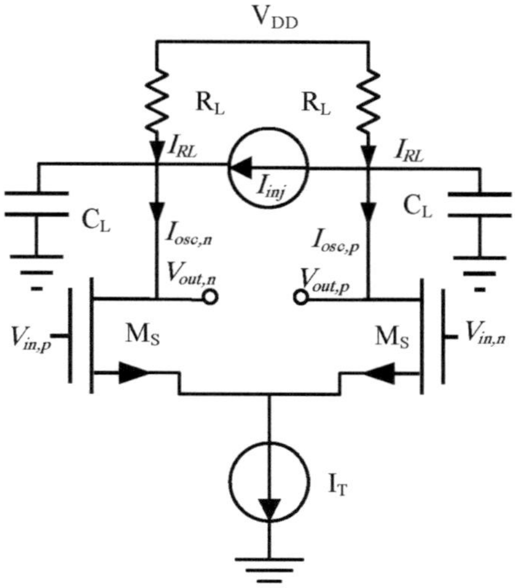

(a)

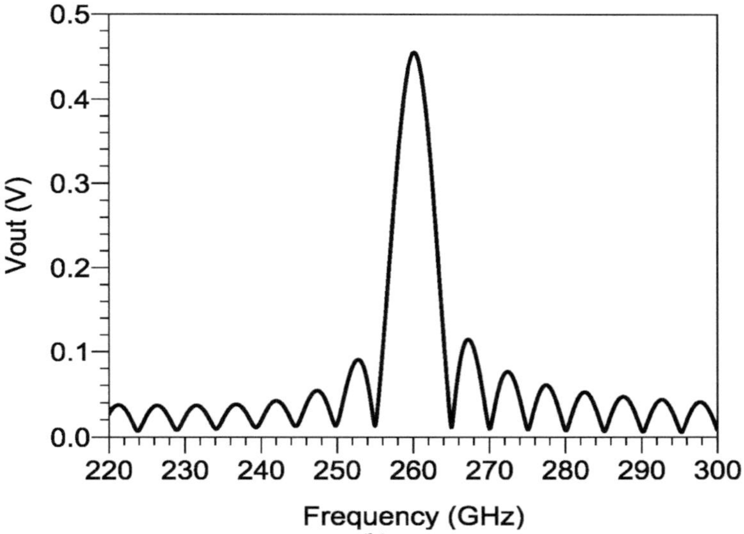

(b)

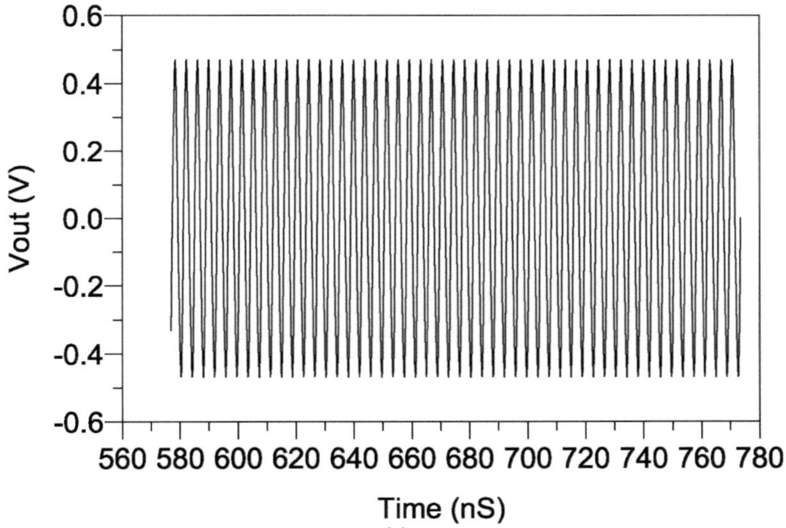

(c)

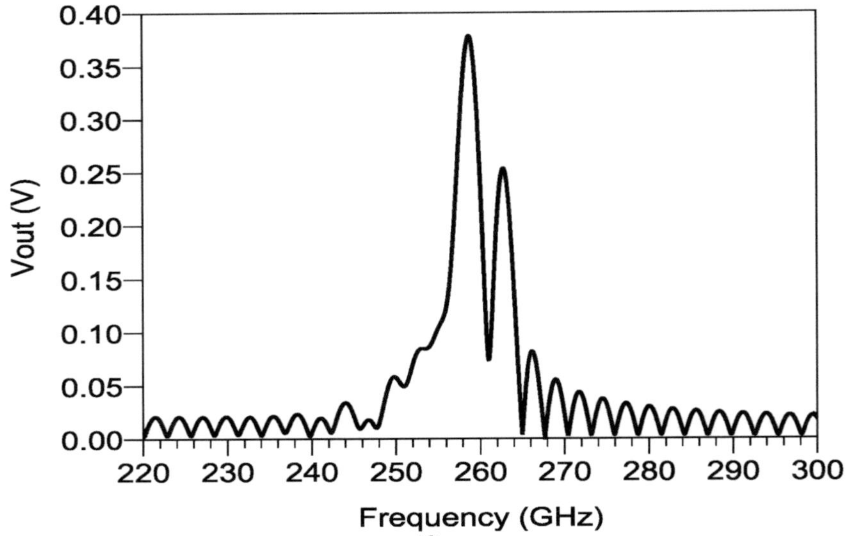
0.40
0.35
0.30
0.25
0.20
0.15
0.10
0.05
0.00
Vout (V)
220 230 240 250 260 270 280 290 300
Frequency (GHz)

(d)

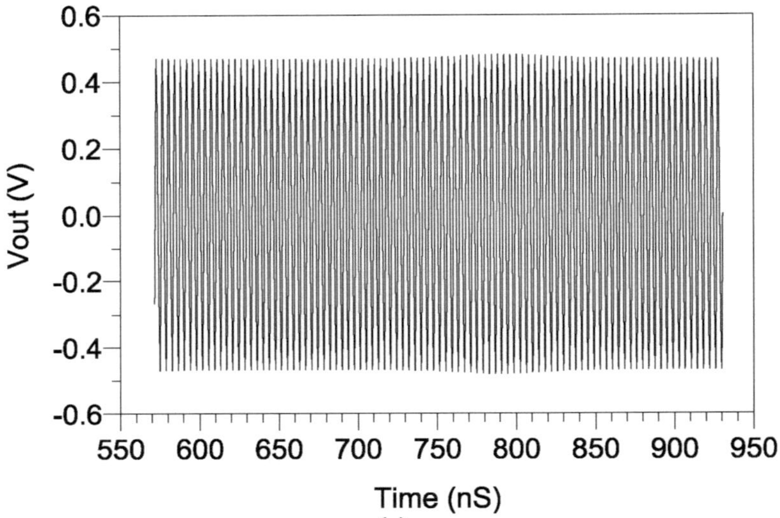
0.6
0.4
0.2
0.0
-0.2
-0.4
-0.6
Vout (V)
550 600 650 700 750 800 850 900 950
Time (nS)

(e)

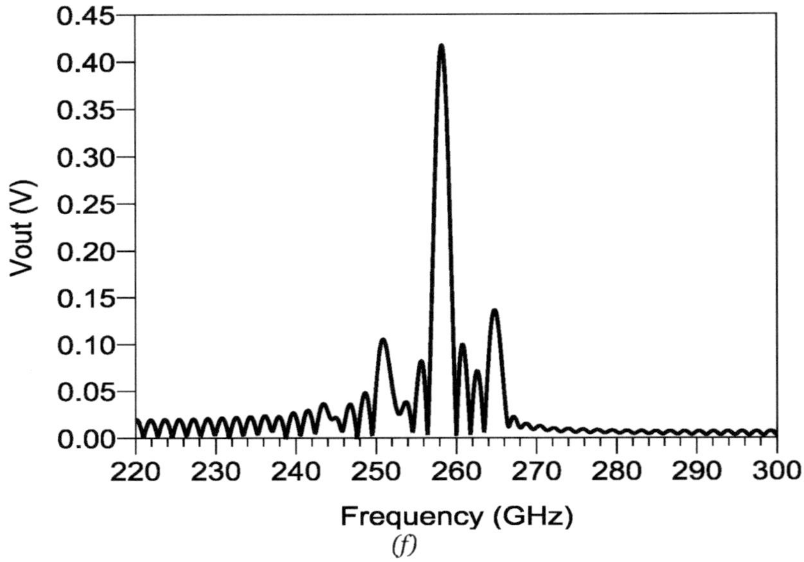
0.45
0.40
0.35
0.30
0.25
0.20
0.15
0.10
0.05
0.00
Vout (V)
220 230 240 250 260 270 280 290 300
Frequency (GHz)

(f)

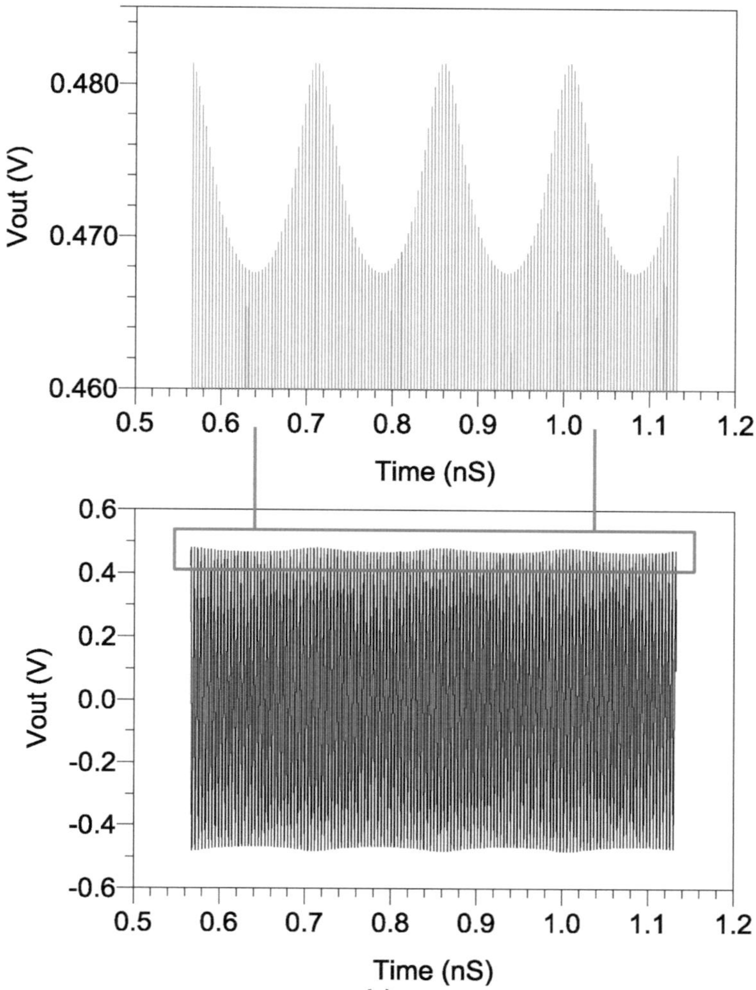

(g)

Figure 2.
(a) Injected delay cell, (b) locking phenomenon in frequency domain (Finj = 260 *MHz and* $I_{inj} = I_{osc}/20$), *(c) locking phenomenon in time domain* (Finj = 260 *MHz and* $I_{inj} = I_{osc}/20$), *(d) pulling phenomenon in frequency domain* (Finj = 262.5 *MHz and* $I_{inj} = I_{osc}/20$), *(e) pulling phenomenon in time domain* (Finj = 262.5 *MHz and* $I_{inj} = I_{osc}/20$), *(f) pulling phenomenon in frequency domain* (Finj = 265 *MHz and* $I_{inj} = I_{osc}/20$), *(g) pulling phenomenon in the time domain* (Finj = 265 *MHz and* $I_{inj} = I_{osc}/20$).

operated similarly to a band-pass filter, and harmonic oscillators have a low-pass filter. The pulling and locking equations are different in these classes. Most of the methods can be used for a class. The locking phenomena have been detected in pendulum clocks in 1629–1695. At first, Adler explained pulling and locking phenomena for LC oscillators for a weak injection signal [9, 54]. Then, case studies that are more special have been reported such as [8, 10–17]. In [10], Adler's equation was extended for applications that the strength of the injection signal is not weak. In [8], by vector diagram of the instantaneous currents, the locking range (LR) equation was improved. The same locking range equation was obtained by another method in [16] and partly in [24, 29]. In [22] using nonlinear feedback analysis, injection locking was investigated. A cross-coupled oscillator under the injection signal is depicted in **Figure 3**. By writing differential equations on the output nodes of the oscillator and solving them in nonlinear, vector diagram or empirical methods, the locking range of the cross-coupled oscillator is achieved.

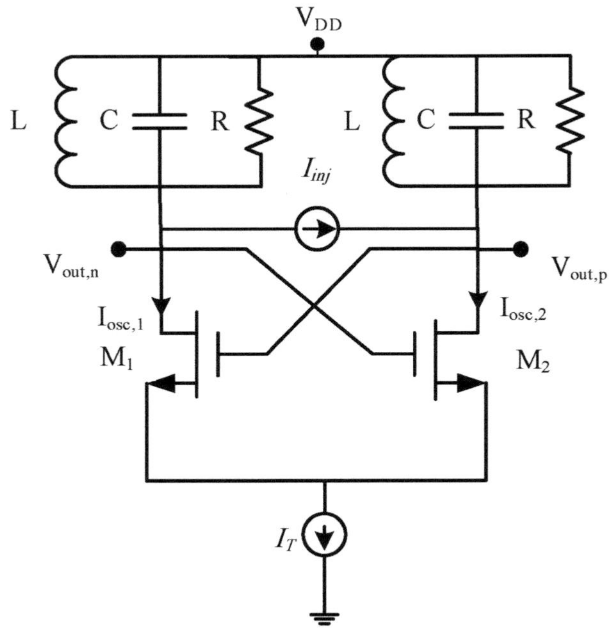

Figure 3.
Cross-coupled LC oscillator.

In **Table 1**, previously significantly published locking ranges are offered, which Q is the quality factor of the LC tank circuit. When $I_{inj}/I_{osc} \ll 1$, it is clear that the locking range is proportional to I_{inj} and inversely proportional Q and I_{osc}.

The pulling phenomena for LC oscillators were completely investigated in [8, 25, 27, 28, 33, 39, 41, 47–49]. When $I_{inj}/I_{osc} \ll 1$, the beat frequency mathematical formula (ω_b) for LC oscillators is as follows:

$$\omega_b|_{LC} = \sqrt{\Omega^2 - K^2}, K = (\omega_0 I_{inj})/(2QI_{osc}), \Omega = |\omega_{osc} - \omega_{inj}| \quad (1)$$

According to Eq. (1), ω_b is increased by increasing Ω or I_{osc} when other parameters are constant. Eq. (1) was improved for strong injection in [41, 47]. The multi-injection signals with different total phases have been reported for the LC oscillators in [24, 29, 56].

Nonharmonic injected oscillators, for instance, ring oscillators and relaxation oscillators, have been studied in several papers [21, 23, 26, 30, 46, 50, 52]. In [21], by using approach of the LC oscillators, a three-stage single-ended ring oscillator was studied. In [23, 30], a locking range equation was obtained at the time domain analysis. In [26], by using current phase diagrams, a locking range was calculated. The locking range of [26] was improved in [50] for a larger injection level. In [52], various locking range equations for ring oscillators were introduced. In addition, comprehensive and exact analyses of the ring oscillators were presented to both

Refs.	Locking range
[9, 44, 54]	$\omega_{osc} I_{inj}/(2QI_{osc})$
[8, 24, 47]	$\omega_{osc} I_{inj}/\left(2QI_{osc}\sqrt{1-(I_{inj}/I_{osc})^2}\right)$

Table 1.
Previously significant published locking ranges for LC oscillators.

Refs.	Locking range
[5]	$\frac{I_{inj}}{NC_L\sum_{i=1}^{\infty}(-1)^{i+1}(2i-1)A_{2i-1}}$
[23, 30]	$\frac{2\omega_{osc}I_{inj}}{NI_{osc}\sin(2\pi/N)}$
[26]	$\frac{\omega_{osc}}{N}\frac{1+\tan^2(\pi/N)}{\tan(\pi/N)\sqrt{1-\left(\frac{I}{I_{osc}}\right)^2}}\frac{I_{inj}}{I_{osc}}$
[50]	$\frac{\frac{\omega_{osc}}{N}\frac{1+\tan^2(\pi/N)}{\tan(\pi/N)}}{\sqrt{1-\left(\frac{I_{inj}}{I_{osc}}\right)^2}-\tan(\pi/N)\frac{I_{inj}}{I_{osc}}}\frac{I_{inj}}{I_{trans.}}$

A_i *is the amplitude of the* i^th^ *output voltage harmonic and* N *is the number of stages.*

Table 2.
Previously significant published locking ranges for ring oscillators.

pulling case and locking case considering higher-order harmonics. Moreover, the multi-injection signals with the same frequency and different initial phase (δ) have been achieved for the ring oscillators. Like locking range for LC oscillators, the locking range for rings oscillators are calculated by solving nonlinear differential equations in the output nodes and proportional to I_{inj}/I_{osc} and inversely proportional to the number of stages (N) [52]. In **Table 2**, previously significantly published locking ranges are exhibited. Moreover, the beat frequency mathematical formula for ring oscillators is expressed as below [5]:

$$\omega_b\big|_{ring} = \sqrt{\Omega^2 - k^2}, k = I_{inj}/(NA_1C_L), \Omega = \left|\omega_{osc} - \omega_{inj}\right| \tag{2}$$

Furthermore, the relaxation oscillators with injection signals were presented in [46]. In addition, a perturbation projection vector (PPV) was employed for analyzing both harmonic and nonharmonic oscillators [18–20]. However, using PPV is complicated, and most of the time it needs preprocessing by simulators such as Cadence.

4. Review of the frequency multipliers/dividers

At first, fractional frequency generators utilizing regenerative modulation have been introduced by [61]. A block diagram of the injection frequency divider/multiplier is displayed in **Figure 4**. By subharmonic ($\omega_{inj} \approx \omega_0/M$) or super-harmonic ($\omega_{inj} \approx M\omega_0$) injection signals, frequency dividers and multipliers were reported in many papers such as [13, 15, 17], where, M is a positive number. Frequency dividers were explored in [36, 38, 40, 43–45, 49, 60]. Studies on

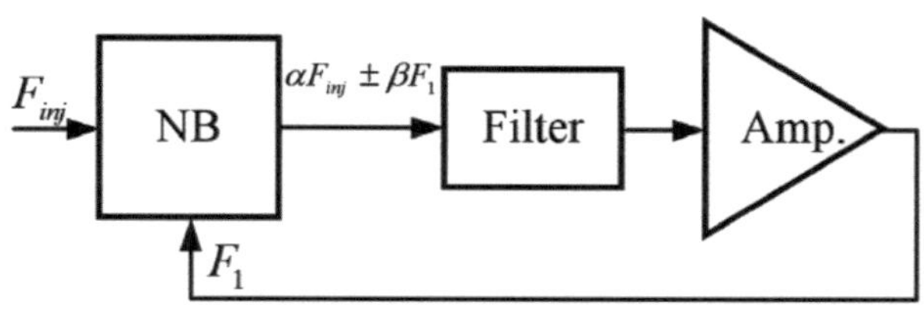

Figure 4.
Block diagram of a frequency divider/multiplier.

frequency multipliers/dividers are generally in two sections. The first one is to obtain an approximation equation of the locking and pulling phenomena [36, 38, 40, 43–45, 49]. The second one is increasing the locking range. In the frequency multipliers/dividers, injection signals may be injected from tail node or output node called direct injection as demonstrated in **Figure 5**. In [13, 15, 17], the general models of the frequency multiplier/dividers have been proposed. The transistor is modeled as a nonlinear block (NB). By using a summer and nonlinear block, a conceptual model was introduced [13]. Nevertheless, once an oscillator behaves similar mixers for the injection signal, for example, when the injection signal is parallel with the tail current source, this model is not correct. In order to achieve a general conceptual model, the summer block is replaced with a multiplier block [17]. However, this model is dependent on SPICE parameters and preprocessing. By phase-domain macromodel, injection-locked frequency dividers were analyzed. Nonetheless, the phase-domain macromodel requests preprocessing and time-consuming. For ÷ 2, an injection signal has been paralleled to the tail current source of the cross-coupled oscillator, and then, the injection signal is modeled as an equivalent injection signal at output nodes whose oscillation frequency and amplitude are $\omega_{inj}/2$ and $2I_{inj}/\pi$, respectively [8]. Therefore, the locking range was

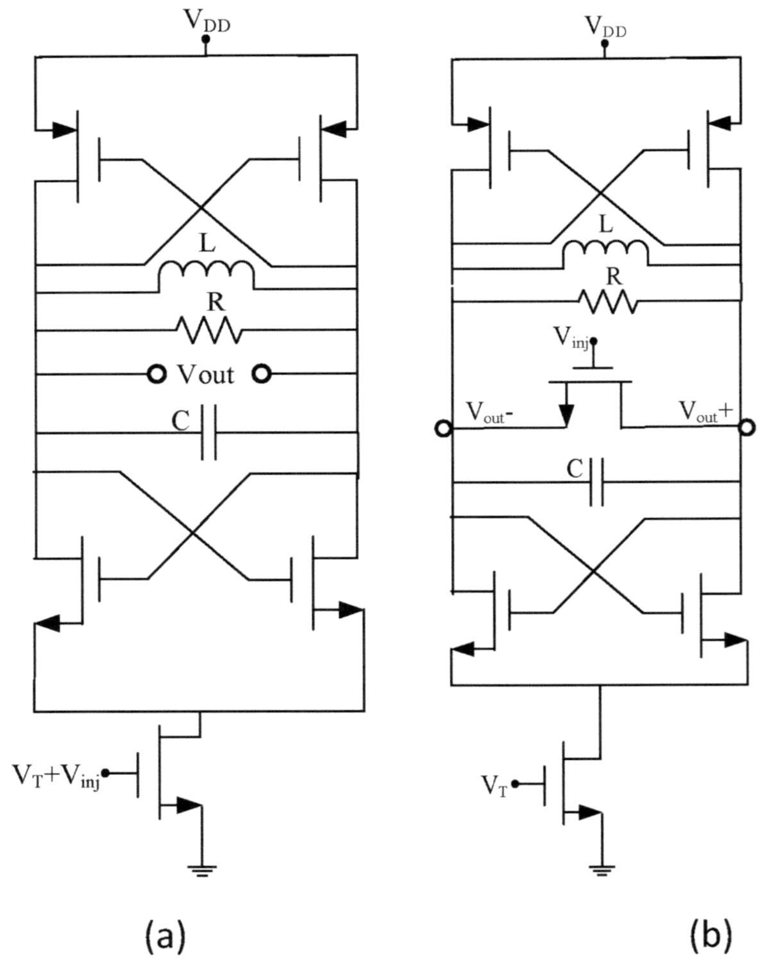

Figure 5.
Conventional injection locking frequency divider/multipliers, (a) injection in the tail, (b) injection in the drain.

DOI: http://dx.doi.org/10.5772/intechopen.91687

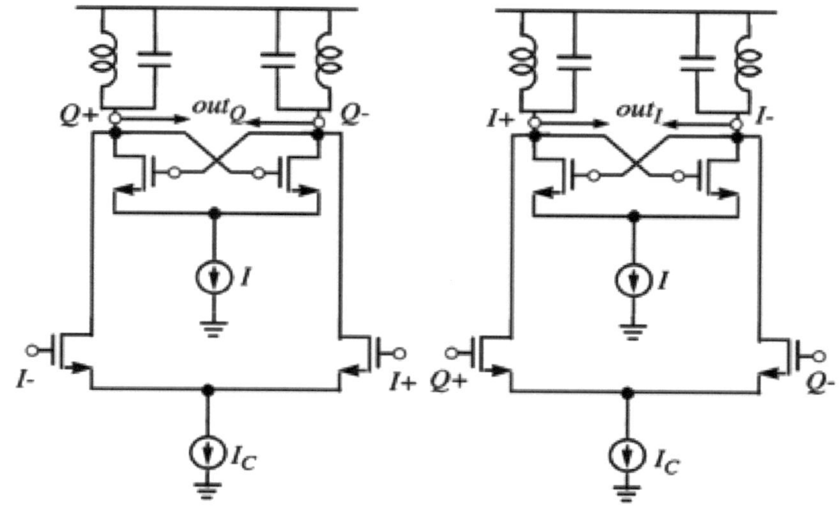

Figure 6.
The quadrature LC oscillator [29].

accomplished similar to the first-harmonic injection locking. This locking range has been acquired by a slightly different analysis in [62]. The asymptotic analysis, or slowly varying amplitude, averaging method, and phase analysis have been utilized to analyze the injected oscillators which make frequency dividers [43–45]. In [45], the locking range has been obtained when the injection signal is applied to the tail current source. Moreover, by the numerical bifurcation analysis using continuation software such as AUTO, they have been analyzed [36, 40]. In [60], an exact analysis for the locking range in injection-locked frequency dividers has been proposed by phasor diagrams and differential equations.

Due to the small locking range of injected LC oscillators, various techniques have been realized to enhance the locking range. Passive and active structures are explored for improving the injection efficiency such as combining inductors in series or parallel with the injection mixer to enhancing its transconductance, body biasing, transformer feedback, dual-resonance RLC resonators, dual injection for increasing the voltage and current injection paths, tapped resonators, switched resonators, harmonic suppression, and distributed injection to distribute the injection signals; in other words, the injection component is divided to several smaller components; input-power-matching and inductive input-matching network is located to the gate of the NMOS switch to heighten the injection power [63–72]. **Figure 6** discloses a quadrature LC oscillator employed in the injection signal.

5. Conclusions

Some basic concepts and definitions have been presented in this chapter. First, pulling and locking phenomena have been introduced which both contain for injection oscillators. Next, previously significantly published papers have been explored. Furthermore, locking range and beat frequency formula have been studied for both first-harmonic injected LC and ring oscillators. Finally, previously significantly published papers about injected locked frequency dividers have been reviewed. Moreover, some previously important published papers about increasing the locking range of the injected locked frequency dividers have been introduced.

Acknowledgments

The *author* would *like* to thank Kermanshah Branch, Islamic Azad University, Kermanshah, Iran, and my wife Dr. Z. Ebrahimipour for any supports.

Author details

Ali Reza Hazeri
Department of Electrical and Electronic Engineering, Kermanshah Branch, Islamic Azad University, Kermanshah, Iran

*Address all correspondence to: alirezahazeri@iauksh.ac.ir

References

[1] Hong B, Hajimiri A. A phasor-based analysis of sinusoidal injection locking in LC and ring oscillators. IEEE Transactions on Circuits and Systems I: Regular Papers. 2018;**66**(1):355-368

[2] Farahabadi PM, Miar-Naimi H, Ebrahimzadeh A. Closed-form analytical equations for amplitude and frequency of high-frequency CMOS ring oscillators. IEEE Transactions on Circuits and Systems I: Regular Papers. 2009;**56**(12):2669-2677

[3] Leung B. VLSI for Wireless Communication. Springer Science & Business Media; 2011

[4] Razavi B. Microelectronics. Upper Saddel River, NJ, USA: Prentice Hall; 2011

[5] Hazeri AR, Miar-Naimi H. Generalized analytical equations for injected ring oscillator with RC-load. IEEE Transactions on Circuits and Systems I: Regular Papers. 2018;**65**(1): 223-234

[6] Hazeri AR, Miar-Naimi H. Novel closed-form equation for oscillation frequency range of differential ring oscillator. Analog Integrated Circuits and Signal Processing. 2018;**96**(1): 117-123

[7] Yuan F. Injection-Locking in Mixed-Mode Signal Processing. Springer; 2020

[8] Razavi B. A study of injection locking and pulling in oscillators. IEEE Journal of Solid-State Circuits. 2004;**39**(9): 1415-1424

[9] Adler R. A study of locking phenomena in oscillators. Proceedings of the IRE. 1946;**34**(6):351-357

[10] Paciorek L. Injection locking of oscillators. Proceedings of the IEEE. 1965;**53**(11):1723-1727

[11] Stover HL. Theoretical explanation for the output spectra of unlocked driven oscillators. Proceedings of the IEEE. 1966;**54**(2):310-311

[12] Campbell C. Beat-frequency spectra in a driven unlocked multimode SAW comb oscillator. In: IEEE 1987 Ultrasonics Symposium. IEEE; 1987. pp. 69-72

[13] Rategh HR, Lee TH. Superharmonic injection-locked frequency dividers. IEEE Journal of Solid-State Circuits. 1999;**34**(6):813-821

[14] Banai A, Farzaneh F. Locked and unlocked behaviour of mutually coupled microwave oscillators. IEE Proceedings - Microwaves Antennas and Propagation. 2000;**147**(1):13-18

[15] Betancourt-Zamora R, Verma S, Lee T. 1-GHz and 2.8-GHz injection-locked ring oscillator prescalers. In: IEEE Symposium on VLSI Circuits, Digest of Technical Papers. 2001. pp. 47-50

[16] Kuo J, Shih E. A 60-GHz 0.13 um CMOS divide-by-three frequency divider. IEEE Transactions on Microwave Theory and Techniques. 2003;**51**(5):1554-1559

[17] Verma S, Rategh HR, Lee TH. A unified model for injection-locked frequency dividers. IEEE Journal of Solid-State Circuits. 2003;**38**(6):1015-1027

[18] Lai X, Roychowdhury J. Capturing oscillator injection locking via nonlinear phase-domain macromodels. IEEE Transactions on Microwave Theory and Techniques. 2004;**52**(9):2251-2261

[19] Lai X, Roychowdhury J. Automated oscillator macromodelling techniques for capturing amplitude variations and injection locking. In: Proceedings of the 2004 IEEE/ACM International Conference on Computer-Aided Design.

IEEE Computer Society; 2004. pp. 687-694

[20] Lai X, Roychowdhury J. Analytical equations for predicting injection locking in LC and ring oscillators. In: Proceedings of the IEEE of Custom Integrated Circuits Conference. IEEE; 2005. pp. 461-464

[21] Mesgarzadeh B, Alvandpour A. A study of injection locking in ring oscillators. In: IEEE International Symposium on Circuits and Systems, ISCAS. IEEE; 2005. pp. 5465-5468

[22] Wan Y, Lai X, Roychowdhury J. Understanding injection locking in negative-resistance LC oscillators intuitively using nonlinear feedback analysis. In: Proceedings of the IEEE of Custom Integrated Circuits Conference. IEEE; 2005. pp. 729-732

[23] Gangasani GR, Kinget PR. A time-domain model for predicting the injection locking bandwidth of nonharmonic oscillators. IEEE Transactions on Circuits and Systems II: Express Briefs. 2006;**53**(10):1035-1038

[24] Mirzaei A, Heidari ME, Abidi AA. Analysis of oscillators locked by large injection signals: Generalized Adler's equation and geometrical interpretation. In: Custom Integrated Circuits Conference. CICC'06. IEEE; 2006. pp. 737-740

[25] Razavi B. Mutual injection pulling between oscillators. In: Custom Integrated Circuits Conference. CICC'06. IEEE; 2006. pp. 675-678

[26] Chien J-C, Lu L-H. Analysis and design of wideband injection-locked ring oscillators with multiple-input injection. IEEE Journal of Solid-State Circuits. 2007;**42**(9):1906-1915

[27] Heidari ME, Abidi A. Behavioral models of frequency pulling in oscillators. In: IEEE International of Behavioral Modeling and Simulation Workshop. BMAS. IEEE; 2007. pp. 100-104

[28] Maffezzoni P, Codecasa L, D'Amore D, Santomauro M. Closed-form expression of frequency pulling in unlocked-driven nonlinear oscillators. In: 18th European Conference on Circuit Theory and Design. ECCTD. IEEE; 2007. pp. 914-917

[29] Mirzaei A, Heidari ME, Bagheri R, Chehrazi S, Abidi AA. The quadrature LC oscillator: A complete portrait based on injection locking. IEEE Journal of Solid-State Circuits. 2007;**42**(9): 1916-1932

[30] Gangasani GR, Kinget PR. Time-domain model for injection locking in nonharmonic oscillators. IEEE Transactions on Circuits and Systems I: Regular Papers. 2008;**55**(6):1648-1658

[31] Bhansali P, Roychowdhury J. Gen-Adler: The generalized Adler's equation for injection locking analysis in oscillators. In: Proceedings of the 2009 Asia and South Pacific Design Automation Conference. IEEE Press; 2009. pp. 522-527

[32] Harutyunyan D, Rommes J, Ter Maten J, Schilders W. Simulation of mutually coupled oscillators using nonlinear phase macromodels. IEEE Transactions on Computer-Aided Design of Integrated Circuits and Systems. 2009;**28**(10):1456-1466

[33] Maffezzoni P, D'Amore D. Evaluating pulling effects in oscillators due to small-signal injection. IEEE Transactions on Computer-Aided Design of Integrated Circuits and Systems. 2009;**28**(1):22-31

[34] Shekhar S et al. Strong injection locking in low-QLC oscillators: Modeling and application in a forwarded-clock I/O receiver. IEEE Transactions on Circuits and Systems I: Regular Papers. 2009;**56**(8):1818-1829

[35] Dal Toso S. Analysis and Design of Injection-Locked Building Blocks for RF Frequncy Generation in Ultra-Scaled CMOS Technologies. PhD [Thesis]. 2010. Available from: http://paduaresearch.cab.unipd.it/3151/

[36] Daneshgar S, De Feo O, Kennedy MP. Observations concerning the locking range in a complementary differential LC injection-locked frequency divider—Part I: Qualitative analysis. IEEE Transactions on Circuits and Systems I: Regular Papers. 2010; **57**(1):179-188

[37] Maffezzoni P. Synchronization analysis of two weakly coupled oscillators through a PPV macromodel. IEEE Transactions on Circuits and Systems I: Regular Papers. 2010;**57**(3):654-663

[38] Maffezzoni P, D'Amore D, Daneshgar S, Kennedy MP. Analysis and design of injection-locked frequency dividers by means of a phase-domain macromodel. IEEE Transactions on Circuits and Systems I: Regular Papers. 2010;**57**(11):2956-2966

[39] Mirzaei A, Abidi AA. The spectrum of a noisy free-running oscillator explained by random frequency pulling. IEEE Transactions on Circuits and Systems I: Regular Papers. 2010;**57**(3):642-653

[40] Daneshgar S, De Feo O, Kennedy MP. Observations concerning the locking range in a complementary differential LC injection-locked frequency divider—Part II: Design methodology. IEEE Transactions on Circuits and Systems I: Regular Papers. 2011;**58**(4):765-776

[41] Ali I, Banerjee A, Mukherjee A, Biswas B. Study of injection locking with amplitude perturbation and its effect on pulling of oscillator. IEEE Transactions on Circuits and Systems I: Regular Papers. 2012;**59**(1):137-147

[42] Yeh Y-L, Chang H-Y. Design and analysis of a W-band divide-by-three injection-locked frequency divider using second harmonic enhancement technique. IEEE Transactions on Microwave Theory and Techniques. 2012;**60**(6):1617-1625

[43] Buonomo A, Schiavo AL. A deep investigation of the synchronization mechanisms in LC-CMOS frequency dividers. IEEE Transactions on Circuits and Systems I: Regular Papers. 2013; **60**(11):2857-2866

[44] Buonomo A, Schiavo AL. Analytical approach to the study of injection-locked frequency dividers. IEEE Transactions on Circuits and Systems I: Regular Papers. 2013;**60**(1):51-62

[45] Buonomo A, Schiavo AL, Awan MA, Asghar MS, Kennedy MP. A CMOS injection-locked frequency divider optimized for divide-by-two and divide-by-three operation. IEEE Transactions on Circuits and Systems I: Regular Papers. 2013;**60**(12):3126-3135

[46] Yuan F, Zhou Y. Frequency-domain study of lock range of non-harmonic oscillators with multiple multi-tone injections. IEEE Transactions on Circuits and Systems I: Regular Papers. 2013;**60**(6):1395-1406

[47] Ali I, Biswas B, Ray S. Improved closed form large injection perturbation analytical model on the output spectrum of unlocked driven oscillator—Part I: Phase perturbation. IEEE Transactions on Circuits and Systems I: Regular Papers. 2014;**61**(1):106-119

[48] Mirzaei A, Darabi H. Mutual pulling between two oscillators. IEEE Journal of Solid-State Circuits. 2014;**49**(2):360-372

[49] Buonomo A, Lo Schiavo A. Evaluating the spectrum of periodic pulling in subharmonic resonant LC circuits. International Journal of Circuit Theory and Applications. 2015;**43**(12):1899-1913

[50] Tofangdarzade A, Jalali A. An efficient method to analyze lock range

in ring oscillators with multiple injections. IEEE Transactions on Circuits and Systems II: Express Briefs. 2015;**62**(11):1013-1017

[51] Ghonoodi H, Miar-Naimi H, Gholami M. Analysis of frequency and amplitude in CMOS differential ring oscillators. Integration, the VLSI Journal. 2016;**52**:253-259

[52] Hazeri AR, Miar-Naimi H. Generalized analytical equations for injected ring oscillator with RC-load. In: IEEE Transactions on Circuits and Systems I: Regular Papers. 2018;**65**(1): 223-234

[53] Imani A, Hashemi H. Distributed injection-locked frequency dividers. IEEE Journal of Solid-State Circuits. 2017;**52**(8):2083-2093

[54] Adler R. A study of locking phenomena in oscillators. Proceedings of the IEEE. 1973;**61**(10):1380-1385

[55] Cheng J-H, Tsai J-H, Huang T-W. Design of a 90.9% locking range injection-locked frequency divider with device ratio optimization in 90-nm CMOS. IEEE Transactions on Microwave Theory and Techniques. 2017;**65**(1):187-197

[56] Tofangdarzade A, Tofangdarzade A, Saniei N. Strong injection locking and pulling in LC multiphase oscillators with multiple injection signals. In: IEEE Transactions on Circuits and Systems II: Express Briefs. 2018;**66**(8):1336-1340

[57] Hong B, Hajimiri A. A Phasor-based analysis of sinusoidal injection locking in LC and ring oscillators. IEEE Transactions on Circuits and Systems I: Regular Papers. 2018;**66**(1):355-368

[58] Hong B, Hajimiri A. A general theory of injection locking and pulling in electrical oscillators—Part I: Time-synchronous modeling and injection waveform design. IEEE Journal of Solid-State Circuits. 2019;**54**(8):2109-2121

[59] Hong B, Hajimiri A. A general theory of injection locking and pulling in electrical oscillators—Part II: Amplitude modulation in LC oscillators, transient behavior, and frequency division. IEEE Journal of Solid-State Circuits. 2019;**54**(8):2122-2139

[60] Mohammadjany A, Hazeri AR, Miar-Naimi H. Exact analyses for locking range in injection-locked frequency dividers. Integration. 2018; **63**:93-100

[61] Miller R. Fractional-frequency generators utilizing regenerative modulation. Proceedings of the IRE. 1939;**27**(7):446-457

[62] Tang-Nian L, Shuen-Yin B, Chen Y. A 60-GHz 0.13 μm CMOS divide by-three frequency divider. IEEE Transactions on Microwave Theory and Techniques. 2008;**56**(11):2409-2415

[63] Jang SL, Chang YT, Hsue CW, Juang MH. Wide-locking range divide-by-4 injection-locked frequency divider using injection MOSFET DC-biased above threshold region. International Journal of Circuit Theory and Applications. 2016;**44**(5):968-976

[64] Jang SL, Lin GY. Wide-locking range single-injection divide-by-3 injection-locked frequency divider. Microwave and Optical Technology Letters. 2015;**57**(12):2720-2723

[65] Jang S-L, Chang C-W. A 90 nm CMOS LC-tank divide-by-3 injection-locked frequency divider with record locking range. IEEE Microwave and Wireless Components Letters. 2010; **20**(4):229-231

[66] Jang S-L, Cheng W-C, Hsue C-W. Wide-locking range divide-by-3 injection-locked frequency divider using

sixth-order RLC resonator. IEEE Transactions on Very Large Scale Integration (VLSI) Systems. 2016;**24**(7): 2598-2602

[67] Lee SH, Jang SL, Lee CF, Juang MH. Wide locking range divide-by-4 injection locked frequency dividers. Microwave and Optical Technology Letters. 2007;**49**(7):1533-1536

[68] Jang SL, Han JC, Lee CF, Huang JF. A small die area and wide locking range CMOS frequency divider. Microwave and Optical Technology Letters. 2008; **50**(2):541-544

[69] Jang SL, Liu CC. Wide-locking range divide-by-4 injection-locked frequency dividers. Microwave and Optical Technology Letters. 2008; **50**(12):3229-3232

[70] Chen CZ, Hsu WL, Lin YS. A 58-GHz wide-locking range CMOS direct injection-locked frequency divider using input-power-matching technique. Microwave and Optical Technology Letters. 2009;**51**(3):685-689

[71] Jang SL, Chang CW, Yang SM. Low power wide-locking range CMOS quadrature injection-locked frequency divider. Microwave and Optical Technology Letters. 2009;**51**(10): 2420-2423

[72] Jang SL, Huang JF, Lin FB. Wide-locking range LC-tank divide-by-4 injection-locked frequency divider using transformer feedback. International Journal of RF and Microwave Computer-Aided Engineering. 2015; **25**(7):557-562

Chapter 7

A Review: Circuit Theory of Microstrip Antennas for Dual-, Multi-, and Ultra-Widebands

Ashish Singh, Krishnananda Shet, Durga Prasad, Akhilesh Kumar Pandey and Mohammad Aneesh

Abstract

In this chapter, a review has been presented on dual-band, multiband, and ultra-wideband (UWB). This review has been classified according to antenna feeding and loading of antennas using slots and notch and coplanar structure. Thereafter a comparison of dual-band, multiband, and ultra-wideband antenna has been presented. The basic geometry of patch antenna has been present along with its equivalent circuit diagram. It has been observed that patch antenna geometry for ultra-wideband is difficult to achieve with normal structure. Ultra-wideband antennas are achieved with two or more techniques; mostly UWB antennas are achieved from coplaner structures.

Keywords: dual-band, multiband, ultra-wideband, microstrip patch

1. Introduction

Antenna is a device that is used to transmit and receive the information in the form of electromagnetic waves only. Antenna is generally classified according to the frequencies, low-frequency, medium-frequency, and high-frequency antenna. Firstly, low-frequency antenna was proposed by a German physicist H. Hertz, i.e., a dipole antenna; thereafter it took 20 years to practically install this antenna. Marconi preformed an experiment on half-wave dipole to transmit the information from Atlantics and received successfully at the receiver side, i.e., St. John's Newfoundland. It was Marconi who used the theory of Prof. J.C. Bose for successful transmission of information. Later, J.C. Bose was known for his contributions on horn antennas. These transmissions of signal were limited to a 200-m distance, and after that De Forester and Felleming developed vacuum tubes; this leads to the increase in transmission distance till 600 m. This enhances the role of electrical and electronics devices for long-distance communications. Then medium-frequency antenna ranges from 300 to 3000 KHz which was designed in the late 1920s. This antenna was constructed using steel bar supported via wooden bars. Later on all lots of development gone till the 1960s for the improvement of medium-frequency antenna, and various antennas were proposed that satisfied this ranges. High-frequency antennas range from 0.003 to 40 GHz after that it is millimeter waves. High-frequency antennas are aperture antenna (pyramidal horn, circular horn, rectangular waveguide), antenna array (linear or planar arrays) and reflector antennas (parabolic reflectors with horn fed or Cassegrain feed, lens antenna, and microstrip antennas (MSAs)) which comprise of a circular, rectangular, or square conducting patch that is made of grounded substrate. This chapter is focused on the miniaturized high-frequency devices so that it can be used in compact wireless communication devices.

In view of this, high-frequency patch antennas devices are introduced, and an extensive literature survey is performed. High-frequency MSAs are the antennas in which dielectric substrates are between the radiating patch and ground plane [1] and are shown in **Figure 1** along with its equivalent circuit diagram. Patch antenna at high frequency is represented as a parallel combination resistance R_1, capacitance C_1, and inductance L_1. The value and equations of C_1, R_1, and L_1 [1] vary depending

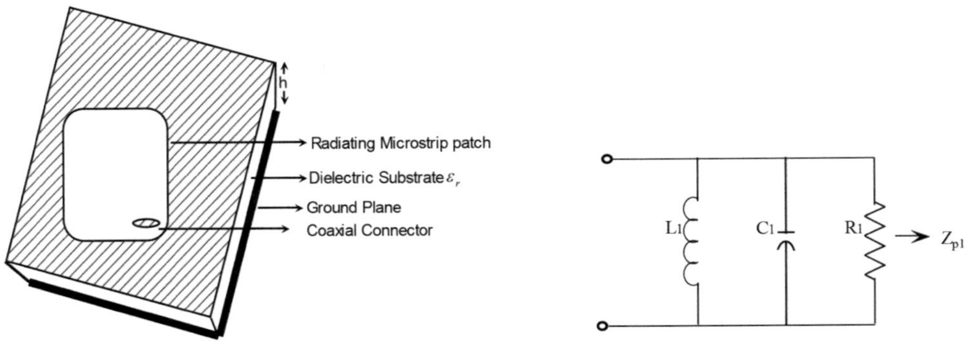

Figure 1.
Configuration of microstrip antenna and its equivalent circuit diagram.

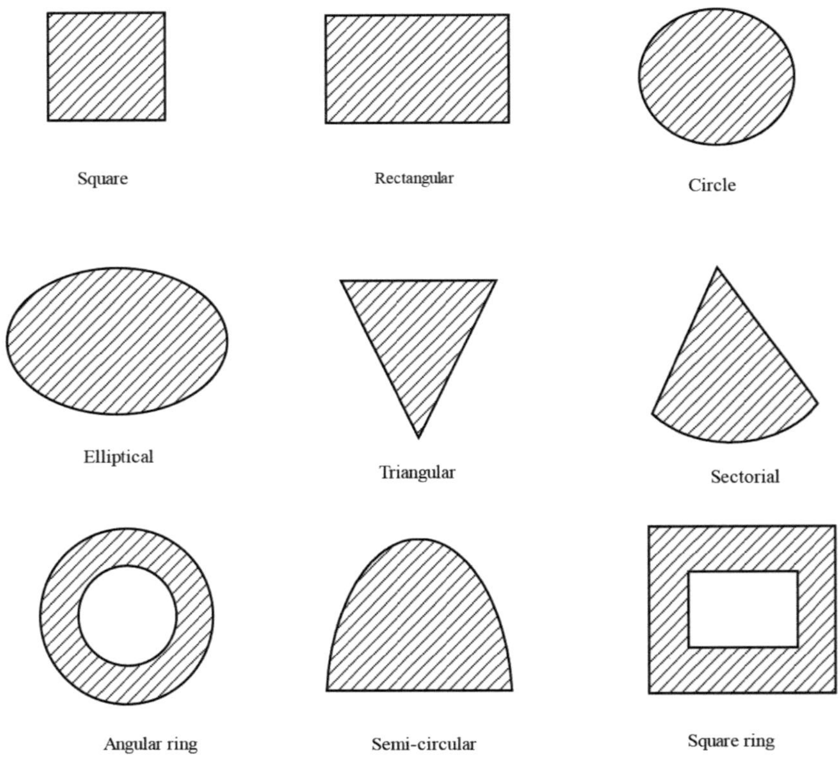

Figure 2.
Different shapes of radiating patches.

on the shape and size of antennas. Patch antennas can be of any shape [1] and size, as shown in **Figure 2**. **Figure 2** shows different types of existing geometry of antennas such as square, rectangular, circle, elliptical, triangular, sectorial, angular ring, semicircular, and square ring. First microstrip patch antenna was reported by Deschamp [2] in 1953. High-frequency microstrip patch antennas generally are divided according to the frequency bands, i.e., dual-band, multiband, broadband, and ultra-wideband (UWB) [3–5]. To achieve these frequency bands, antennas of different shapes and sizes have been presented. In the next section, these frequency bands are discussed with respect to their shape, size, and feeding techniques. From **Figure 1** the input impedance of patch can be written from its equivalent circuit as

$$Z_{p1} = \frac{1}{\frac{1}{j\omega L_1} + j\omega C_1 + \frac{1}{R_1}} = \frac{j\omega L_2 R_1}{j\omega L_2 - \omega^2 L_1 C_1 R_1 + R_1}, \tag{1}$$

2. Dual-band MSAs

Dual-band and multiband microstrip patch antenna can be realized with different techniques such as stacking, coplanar structures (parasitic patches), slots, notches, shorting pin, shorting wall and active devices, etc. Using these techniques several research papers were published by various researchers, and the first dual-band radiator was reported in 1984, using shorting pins in rectangular patches by Wang and Lo [6]. In this section, literature survey of dual-band and multiband is presented on the basis of techniques used to achieve dual or multiband.

2.1 Slot loaded MSAs

Slot loaded MSAs can be achieved by etching rectangular or slot of any desired shape in the patch as shown in **Figure 3**. This antenna can be represented as a parallel combination of impedance due to patch (Z_{p1}) and impedance due to slot (Z_{sh}), and its equivalent is denoted as (Z_{ps}), as given in **Figure 3** and calculated as,

$$\frac{1}{Z_{ps}} = \frac{1}{Z_{p1}} + \frac{1}{Z_{sh}}, \tag{2}$$

The slotted loaded MSAs proposed by various researchers are summarized here.

Daniel and Shevgoankar [7] studied rectangular microstrip patch antenna with the slot etched along with the nonradiating edge of the patch antenna for tunable dual-band operation and investigated the effects of the slot parameter on the tuneableness of the RMSA resonant frequencies. Wu [8] demonstrated compact slot loaded triangular patch antenna for wireless application and measured gain as 2.8

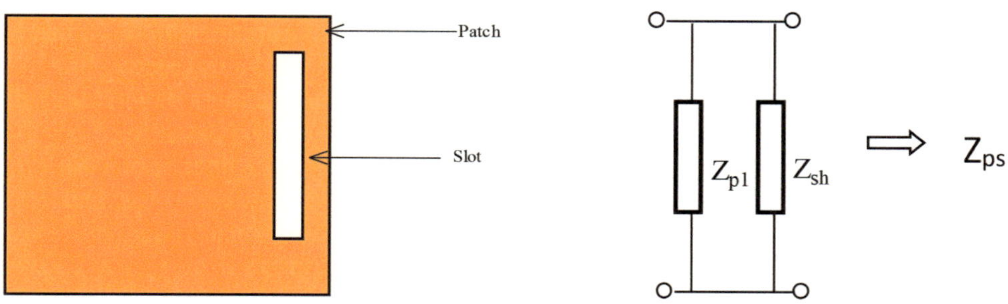

Figure 3.
Top view of slot loaded MSAs and its equivalent circuit diagram.

and 5 dBi with variations within 1.0 dBi. Vishvakarma and Vishvakarma [9] presented the theoretical analysis of inclined slot loaded rectangular patch antenna for dual-band operation. They given circuit diagram for patch antenna which is similar to **Figure 3** and circuit theory concept was used to calculate reflection loss, variation of slot length and width was also discussed. Lim et al. [10] presented the modeling of dual-band aperture-coupled microstrip patch antenna (ACMPA) with a symmetric crossed slots. Ansari et al. [11] also reported a triple U-shaped slot loaded patch antenna for Bluetooth and wireless local area network (WLAN) application and have explained that by introducing three U-slots in the circular disk, 16.54% of size reduction can be achieved. Lu and Liu [12] proposed a novel planar slot array antenna with high-gain operation for long-term evolution (LTE)/Worldwide Interoperability for Microwave Access (WiMAX) point and measured maximum peak antenna gains and efficiencies of 13.9/14.1 dBi and 86.5/73.5% across 2.6/3.5 GHz bands, respectively, whereas Luo et al. [13] reported an open L-slot antenna for WLAN/WiMAX applications. Ansari et al. [14] reported single W-slot loaded and two dielectric layers microstrip patch antenna using equivalent circuit theory concept and reported the frequency ratio and 1.33 for single-layer patch antenna and 1.38 for two-layer patch antenna. Gupta et al. [15] proposed a microstrip antenna with tunable high-impedance surface with two varactor diodes. Mohammad et al. [16] proposed a U-shaped patch, T-shaped monopole path, and a pentagonal wide slot in the ground plane for multi-input multi-output (MIMO)/diversity applications. Chakraborty et al. [17] designed two different single slotted loaded on rectangular microstrip antennas with slotted ground plane for IEEE 802.11a WLAN application. Tsai [18] investigated a bow tie-shaped CPW-fed slot antenna which consists of a coaxial connector and two conducting strips for wireless communication applications.

2.2 Feeding techniques of MSAs

There are different feeding techniques by which dual-band can be achieved such as coplanar waveguide (CPW) fed, coaxial fed, microstrip line-fed, etc. and is shown in **Figure 4**. **Figure 4(a)** shows the CPW fed in which ground plane and excited patch lies on the same plane. **Figure 4(b)** shows microstrip line fed of 50 Ω which is used to excite the patch and ground plane that is on either side of the dielectric substrate. **Figure 4(c)** shows coaxial fed in which coaxial fed feed from bottom of antenna via drilling hole till excited patch. **Figure 4(d)** shows proximity fed which has actually microstip line fed which lies between two dielectric substrate of height H_1 and H_2. **Figure 4(e)** it has microstrip line fed at bottom of the substrate above that ground plane at height H_2 is placed above the ground plane patch is placed with height of the dielectric substrate H_1. Different feeding techniques used by various researchers to achieve dual-band are briefly described below.

Liu and Chen [19] proposed a compact patch antenna CPW fed for dual-band operation and obtained that lower band and upper band of antenna can be utilized for short-range wireless application. Sung et al. [20] illustrated a microstrip line dual frequency dielectric resonator antenna printed on a single metal layer. Su et al. [21] presented a rectangular patch antenna surrounded by a ring patch for dual-band operation and observed that two operating bands produce linear polarization. Chen and Hsu [22] proposed broadband radial slot CPW-fed antennas for dual operation and observed that the bandwidth of the lower resonance frequency lies in the range of 2.4–9.7%, while that of the upper frequency ranges from 17.4 to 23.2%. Sze et al. [23] proposed a circular slot antenna fed by a coplanar wave guide with slit back patch, and they observed that broadside far-field pattern at both frequency bands. Huang et al. [24] studied dual-band monopole antenna excited by a

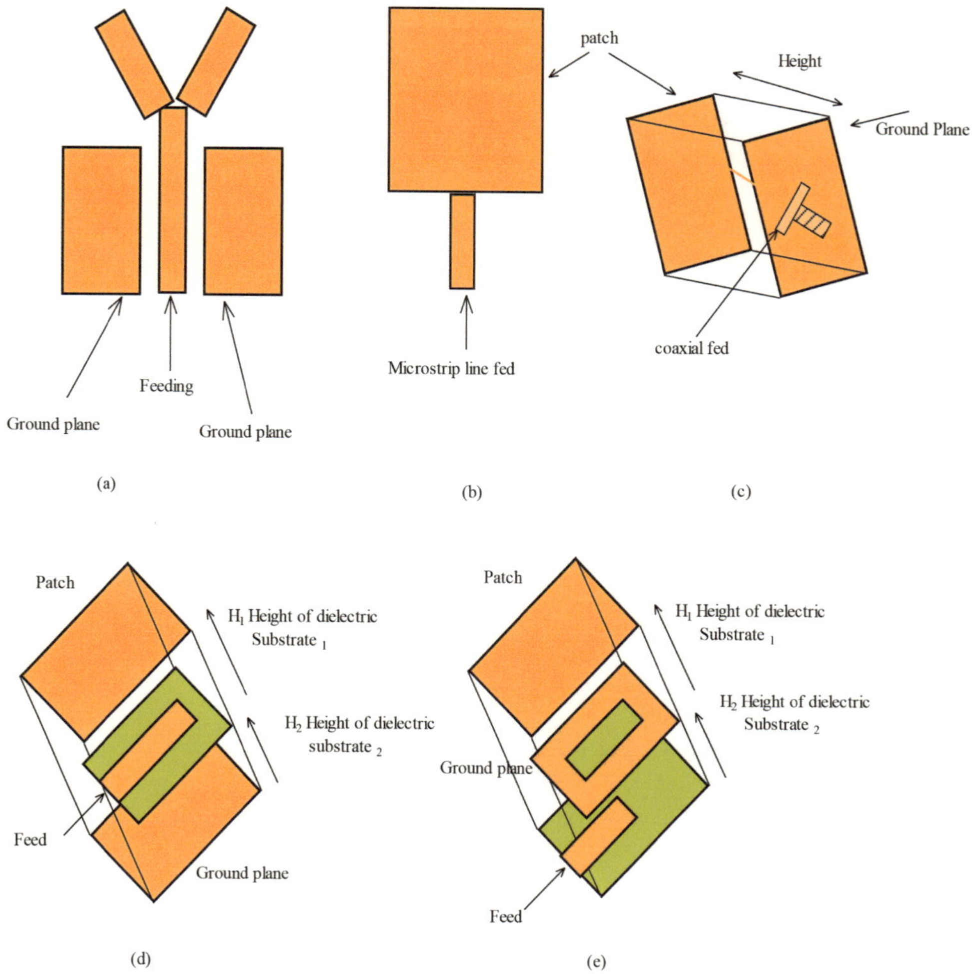

Figure 4.
Different feeding techniques of MSAs. (a) CPW fed, (b) microstrip line fed, (c) coaxial fed, (d) proximity fed, and (e) aperture fed.

capacitive coupling feed for WLAN application and measured 3.4% and 13% bandwidth for upper and lower resonance frequency. Yoon et al. [25] designed a planar CPW-fed slot antenna and thin substrate for dual-band operation for WLAN applications and measured peak gains 1.37 and 1.7 dBi for lower and upper frequencies, respectively. Lin et al. [26] investigated a compact CPW-fed patch antenna with two embedded slots and found average antenna gain as 1.4 and 5.1 dBi.

Singh et al. [27] published an article on aperture-coupled feeding to achieve dual-band. The antenna was designed in the IE3D simulation software, and results were compared with theoretical result via circuit theory concept. Two resonance frequencies were achieved at 4.39 and 5.55 GHz with a percentage bandwidth of 10.23 and 13.33%. Circuit concept was applied on antenna for analysis of these antenna parameters such as input impedance, reflection coefficient, VSWR, and return loss. Antenna physical parameters were also varied such as dielectric substrate thickness of two dielectric used, width and length of aperture, and dielectric constant of two dielectric used in the design for achieving dual-band characteristics.

Microstrip line fed exciting the gap coupled rectangular and parasitic patch was presented by Singh et al. [28]. Dual-band was achieved through spitted ring resonator (SRR) at ground plane and parasitic patch at 0.9 and 1.8 GHz. They proposed the circuit of design of antenna, and analysis of the antenna was presented theoretically on varying gap between parasitic and fed patch, microstrip width and length, and height of substrate. All these variation was explained in terms of change in inductance and capacitance.

Gulam Nabi Alsath et al. [29] discussed and designed two identical aperture-coupled pattern reconfigurable dielectric resonator antennas (DRAs). Li et al. [30] studied a compact asymmetric coplanar strip (ACS)-fed using loaded capacitance terminations for 2.4/5.8 GHz wireless local area network applications. Moosazadeh and Esmati [31] presented the small planar microstrip-fed square radiator using slotted conductor backed plane for wireless local area network applications. Chen et al. [32] investigated the slot stepped-impedance resonators (SIRs) and CPW feeding for dual-band/tri-band/broadband applications. Emadian and Shokouh [33] investigated CPW-fed slot antenna which consists of a coaxial connector and two conducting strips for wireless communication applications with high increased impedance bandwidth.

Singh et al. [34] designed an antenna with L-strip feed. It is a slightly complicated geometry to be fabricated because it is difficult to create an L-strip fed structure. This fed excite rectangular patch with three rectangular slots. They found that resonating frequency depends on the slot length and width, L-strip length, and width which have a vital role in achieving dual-band.

2.3 Stacked MSAs

Stacking of MSAs means having two or more dielectric substrate of different or same thickness that is kept over each other or one can say the toppling of one substrate over the other. Dual-band can also be achieved by stacking as shown in **Figure 5**. From figure, patch is at top of the antenna that is excited through coaxial fed below the excited patch one more patch placed of height H_2 (no-excited) after ground plane is there which is ground through coaxial fed. **Figure 5** equivalent circuit represents combination impedance of excited patch Z_{P1}, and electromagnetic coupling occurs due mutual inductance and capacitance occurs, i.e., mutual inductance L_m, inductance occurs because of patch and other patch L_{mp}, capacitance occurs because of patch and other patch C_{mp}, and L_{p1} and C_{p1} are inductance due to patch and ground plane. Its equivalent impedance can be given as

$$Z_{St} = \frac{1}{Z_{p1}} + \frac{1}{\left(\frac{1}{L_m} + \frac{1}{L_{p1}}\right) + L_{mp}} + \frac{1}{\left(\frac{1}{C_{mp}} + C_{p1}\right) + C_m}, \quad (3)$$

Karmakar [35] proposed shorted strap tunable stacked patched PIFA and measured bandwidth of 10.8% at 1.8 GHz where shorting and stacked structure plays

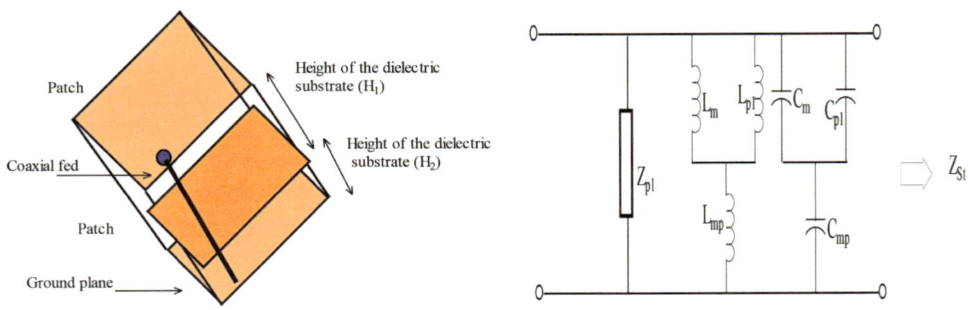

Figure 5.
Side view of stacked MSAs and its equivalent circuit diagram.

a vital role in achieving 10.8% bandwidth. Pioch and Laheurte [36] investigated a dual-band stacked microstrip based on a small parasitic patch inserted slightly above an EBG patch. Kim et al. [37] presented the design simulations and measured stacked meandered patch antenna for mobile communication and observed that dual-band operation is realized using via holes with 0.3-mm height. Anguera et al. [38] presented stacked microstrip patch antenna using reactive loading and a fractal shape radiating edge. Ansari et al. [39] reported the analysis of U-slot loaded patch stacked with H-shaped parasitic elements. The circuit theory concept was applied by them and presented equivalent input impedance of the designed antenna. Further, they calculated variation of U- and H-shaped slotted on antenna bandwidth. Ma and Row [40] proposed a design of single-fed antenna structure composed of two stacked patches with different polarizations and radiation patterns. Benkouda et al. [41] investigated stacked high T_c superconducting rectangular patches fabricated on a two-layered substrate using a full-wave spectral analysis. Batgerel and Eom [42] presented a stacked microstrip patch antenna structure combined with high-gain dielectric rod and a sleeve-dipole element. Mishra et al. [43] designed the dual- and wideband slot loaded stacked microstrip patch antenna for WLAN/WiMAX applications. The bandwidth of the proposed dual-band antenna at lower resonance frequency is 9.53%, whereas at upper resonance frequency, 6.95% is achieved.

2.4 Parasitic MSAs

Parasitic MSAs can be designed as shown in **Figure 6**, with its equivalent circuit diagram. These types of antenna are represented as a combination of gap capacitance (C_{p1} and C_{p2}) with patch and in addition to it develop coupling capacitance C_c, and the values of C_{P1}, C_{P2}, and C_C can be calculated as in [1]. Its equivalent impedance can be given as

$$Z_{CC} = j\omega C_g + j\omega C_{P1} + \frac{1}{j\omega C_{p1}}, \tag{4}$$

It has the fed patch and non-fed patch on the same plane of the dielectric substrate. Brief survey of antennas having parasitic elements is described here.

Anguera et al. [44] reported a dual frequency patch antenna based on the Sierpinski fractal geometry with two parasitic patches to enhance the bandwidth of

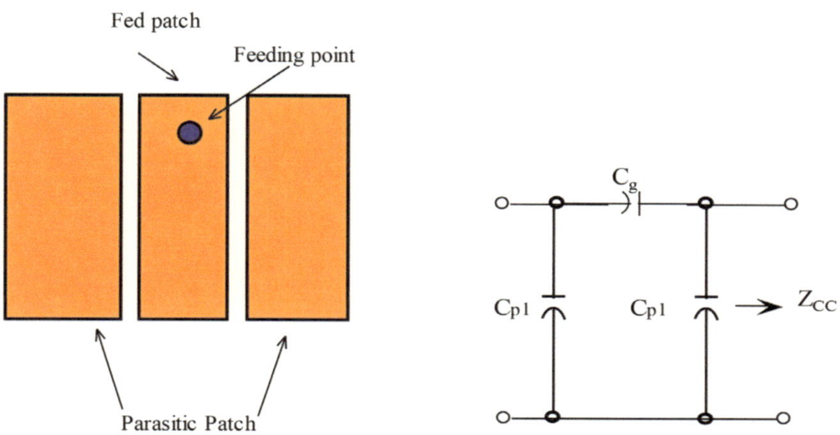

Figure 6.
Top view of parasitic MSAs and its equivalent circuit diagram.

the antenna. Chen et al. [45] proposed planar antenna, an ungrounded dielectric slab with parasitic reflector for dual-frequency operation and steerable end-fire pattern characteristics. Cho et al. [46] presented a compact internal antenna with parasitic patch for mobile handset and measured the bandwidths of 140 MHz (1740–1880) for which is the within the range of Korean communication service (KPCS) band and 90 MHz (2400–2490) in the Bluetooth devices. Liu et al. [47] presented a triangular patch antenna with parasitic element over a modified ground plane and measured bandwidth greater than 20% at the 2.45 and 5 GHz bands. Peng and Ruan [48] presented microstrip-fed dual-band design patch antenna with two parasitic invert L stubs for 2.4/5-GHz wireless applications, and they found L stub plays important part in achieving dual-band.

2.5 Notch-loaded MSAs

Dual-band-notch-loaded MSA geometry is shown in **Figure 7** and its equivalent circuit diagram. The figure shows rectangular patch with coaxial fed having rectangular notch at edges. Due to notches ΔL and ΔC, inductance and capacitance will be developed, and it is represented as a parallel combination of R_2, $L_1 + \Delta L = L_2$, $C_1 + \Delta C = C_2$ which can be calculated [1]. Its equivalent impedance is given as

$$Z_n = \left(\frac{1}{R_2} + \frac{1}{L_1 + \Delta L} + \frac{1}{C_1 + \Delta C}\right) \tag{5}$$

Rectangular and circular notch is etched near the edges of the MSAs as seen in the figure. Depending on the designs of notch-loaded MSAs, brief literature survey is done.

Shivnarayan and Vishvakarma [49] investigated the theoretical analysis of notch-loaded patch antenna for dual-band operation, and they observed that frequency ratio varies from 1.11 to 1.83 with length and width of the notch. Deshmukh and Kumar [50] analyzed various types of dual-band slotted antennas such as U-shape slotted, E-shaped notched, and pair slots. They found the U-shaped and pair of slots help in achieving dual-band and discussed radiation characteristics of the antenna. Mishra et al. [51] presented notch-loaded microstrip patch antenna with shorting pin using cavity model. They achieved dual-band on inserting shorting pin, and variation of notch length and width was also presented. Thomas and Sheernivasan [52] proposed a novel CPW antenna which comprises a rectangular patch and notch cut at lower edge of the patch and measured gain of 2 dBi. A compact-notched CPW-fed wide-slot antenna for WLAN and WiMax

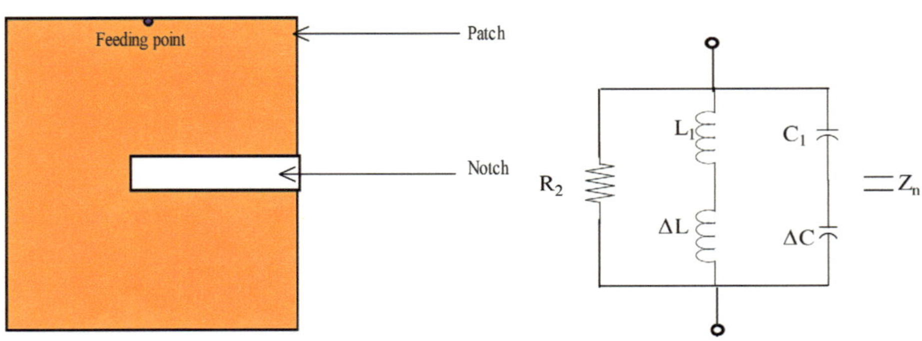

Figure 7.
Top view of notch-loaded MSAs and its equivalent circuit diagram.

applications was proposed by Lin et al. [53]. Singh et al. [54] presented and analyzed two symmetrical notches and shorting pin-loaded patch antenna using circuit theory concept based on modal expansion cavity model. Variation of symmetrical notches' length and width was shown using circuit theory concept, and parametric analysis of antenna design was also presented. Mishra et al. [55] analyzed L-strip-fed circular disk patch antenna using circuit theory concept and observed that when the notch is etched on the circular disk, patch dual-band behavior shifts towards resonance higher side.

Verma et al. [56] designed and fabricated equilateral triangle antenna with edge-notched at each edge of triangle. Dual-band was achieved at 9.53 and 4.81 GHz. The variation of frequency ratio for side length, substrate thickness, and dielectric constant was presented. Antenna was designed for dual-band that is applicable for the C, X, and Ka band operation.

3. Broadband MSAs

Broadband MSAs can be defined as a bandwidth that is greater than 18% for upper and lower resonance frequencies that too band should be −10 dB down. Broadband can be achieved by different techniques such as slots and notches and different feeding techniques such as proximity, aperture, CPW fed, microstrip line, stacked, parasitic elements, etc. In 1979, the first broadband antenna by stacking rectangular patches in two- and three-layered configuration was described by Hall et al. [57]. Long and Walton [3] investigated the stacked microstrip antennas for dual-frequency operation. A brief literature survey of broadband MSAs is presented herein and described according to its feeding techniques, reactive loading, coplanar structures, etc.

3.1 Feeding techniques for broadband MSAs

Mestdagh et al. [58] presented a systematic study of aperture-coupled stacked patch antennas fed by coplanar waveguide useful at millimeter wave frequency in the 30 GHz range. They found the antenna bandwidth of 30%. Stevan et al. [59] proposed broadband patch antenna with L-shaped probe and found bandwidth of 35% with a gain of 6 dBi. They achieved 30% reduction in dimension from convention patch antenna. Yuehe [60] proposed a novel two-layer stacked wideband dielectric resonator antenna (DRA) of rectangular shape. Zhang et al. [61] presented a dielectric resonator antenna formed by a U-shaped dielectric resonator and a conformal elliptical patch proposed for wideband communication applications. The measured impedance bandwidth $S11 < -10$ dB was about 72%, covering the frequency range 3.82–8.12 GHz.

3.2 Stacked design of MSAs

Natrajan et al. [62] introduced a novel topology of a dual-stacked, U-slot loaded patch antenna and achieved VSWR bandwidth of 65% and peak gain of 9.4 dBi. Lui and Luk [63] in 2005 designed a stacked patch antenna fed by meandering probe and achieved 37% bandwidth. Ghannoun et al. [64] reported a patch antenna with two E-shaped stacked patches and found a bandwidth of 54%. Deshmukh and Kumar [65] presented a compact broadband stacked patch antenna and found 10.9% bandwidth for C-shaped stacked patch and 9.7% for H-shaped stacked patch antenna.

3.3 Slot loaded design of MSAs

Wu et al. [66] presents a novel dual-broadband rectangular slot antenna for 2.4 and 5 GHz WLAN. The impedance bandwidth for two operating bands could reach up to 10.6% for the 2.4 GHz band and 33.8% for the 5 GHz band. Deshmukh and Kumar [67] reported a broadband rectangular microstrip antenna with a pair of rectangular. Its bandwidth was further increased either by cutting multiple pairs of slots or by cutting a pair of bowtie slots. Deshmukh and Kumar [68] presented various suspended broadband and compact circular microstrip antennas using variations of U-slot, V-slot, and pair of rectangular slots cut either on the periphery or inside the patch. Fang et al. [69] described a novel microstrip slot antenna backed by a ground plane of large bandwidth and high gain. Chen et al. [70] proposed a novel microstrip line-fed wide-slot antenna. The proposed antenna has a simple symmetric structure and compact size. Mo et al. [71] presented a broadband ultra-high-frequency radio-frequency identification (UHF RFID) tag patch antenna with a pair of U-slots.

Aneesh et al. [72] presented a research article of S-shaped patch antenna on tough form with copper laminated on the front and back side. They applied a multilayer perceptron artificial neural network (MLPANN) for the analysis of antenna design on varying size of notches. The ANN models with several types of algorithm were trained from which Levenberg-Marquardt (LM) algorithm takes less time. S-shaped-type structure is formed by etching two rectangular notches of equal dimension and resonating at 2.64 GHz with a bandwidth of 20.57%.

Aneesh et al. [73] designed an H-shaped rectangular structure on IE3D simulation software and achieved triple band for PCS and WiMAX applications. The design was fed via a microstrip line fed with finite ground plane. They found that measured and simulation results were in close agreement for radiating structure. They observed antenna maximum efficiency and gain at 5.2 GHz, i.e., 95.5% and 7 dBi.

3.4 Other techniques of MSAs

Liu and Boyle [74] proposed a bandwidth enhancement of a quarter wavelength slot antenna by capacitive loading to increase its bandwidth. Tang and He [75] proposed a square microstrip patch with a pair of U and T slots which was suitable for attaching to metallic objects. Kimouche et al. [76] presented the monopole antenna with the bandwidth enhancement technique covering the band 3.1–10.6 GHz, which was approved by the federal communication commission as a commercial band for ultra-wideband communication systems, with a measured return loss bandwidth of 7.9 GHz.

4. Ultra-wideband MSAs

In the year 2002, the Federal Communication Commission (FCC) defined ultra-wideband as those bands which have first resonance frequency at 3.1 GHz and higher resonance frequency at 10.6 GHz. After FCC defined the bandwidth of ultra-wideband, the lots of papers were published on UWB MSAs. Depending on the designs of patch antenna, a brief literature survey is described in the section.

4.1 Slot loaded MSAs for UWB

Liu et al. [77] proposed a microstrip-fed antenna, consisting of a square slot patch with a vertical coupling strip, which only occupies a small size of 15 (L) $\times$ 15

(W) × 1.6 (H) mm^3. Ghaderi and Mohajeri [78] proposed a compact wide-slot antenna with microstrip-fed monopole for UWB application. The monopole is composed of an elliptic patch connected to a trapezoid one. Ojaroudi and Ojaroudi [79] presented a novel inverted T-shaped ring slot antenna, surrounded by a C-shaped slot, which provides a wide usable fractional bandwidth of more than 125% (2.71–12.06 GHz). Peng Gao et al. [80] proposed an inverted U-shaped slot on the ground plane and a radiation patch similar to the slot that is fed by a 50-Ω microstrip line. Shinde and Mishra [81] presented L-shaped thin ground plane for lower frequency band stop function and a tapered rectangular vertical slot in radiating patch for upper frequency band stop. Gong et al. [82] presented a compact ultra-wideband antenna with an L-shaped slot. They found that the proposed antenna offers 118% bandwidth when the antenna is fabricated on dielectric constant 4.4 and of dimension of 23.7 × 23.7 × 0.8 mm^3. Haraz et al. [83] designed two different slot-coupled vertical microstrip-microstrip transitions for UWB multilayer microwave circuits and simulated the results using two simulation programs with two different numerical techniques.

4.2 Notch-loaded MSAs for UWB

Shameena et al. [84] designed and analyzed a compact planar UWB antenna with notch-band ON/OFF control and achieved the UWB response by a microstrip-fed staircase patch with an identical inverted ground plane. Ojaroudi et al. [85] presented a novel printed monopole antenna (PMA) for UWB applications with variable frequency band-notch characteristics. Movahedinia and Azarmanesh [86] presented novel printed monopole antenna for ultra-wideband applications with variable frequency band-notch characteristics. Ojaroudi et al. [87] presented printed monopole antenna with constant gain over a wide bandwidth for ultra-wideband applications with desired notch-band characteristic. Li et al. [88] presented a compact dual-band-notched UWB antenna and showed that the designed antenna is of compact size, has a wide bandwidth covering 3.05–14.2 GHz, realizing dual-notched bands of 5.14–5.36 and 5.74–6.07 GHz. Ojaroudi et al. [89] presented a novel UWB printed monopole antenna for use in a circular cylindrical microwave imaging system. Mehranpour et al. [90] proposed a novel printed monopole antenna for ultra-wideband applications with dual-band-notch function.

Fuguo Zhu et al. [91] proposed novel band-notched antennas suitable for UWB applications with coexisting wireless systems operating over 3.3–3.6, 5.15–5.35, or 5.725–5.825 GHz bands. Mandal and Das [92] presented a printed plaque monopole antenna fed by a microstrip line for UWB width with triple notch band. Natarajamani et al. [93] proposed a compact two-element diversity planar antenna for UWB application with band-notch function. Karimian et al. [94] designed two compact monopole antennas with band-notched characteristics at for ultra-wideband application. Zheng et al. [95] investigated compact optically controlled UWB antenna with reconfigurable band-notched characteristics for cognitive radio (CR) applications. Rajesh Kumar and Raghavan [96] demonstrated a compact microstrip-fed novel U-shaped fractal monopole antenna with enhanced bandwidth and band-notched characteristics for ultra-wideband applications.

Gautam et al. [97] designed UWB antenna for MIMO application. An antenna was excited with microstrip line, and resonating band was observed at 3.1–35 GHz. Within this extended bandwidth, two bands were obtained at 5.1–5.8 GHz and 6.7–7.1 GHz. The variation of antenna design was as presented single and double notches. The center notch frequency F_n was calculated as

$$f_n \approx \frac{c}{4L_f\sqrt{\varepsilon_r + 1/2}}, \tag{6}$$

They also calculated mean effective and diversity gain (DG) and envelope correlation coefficient (ECC) and reflection coefficient (RC)

$$ECC = \frac{|S_{11}^* S_{12} + S_{21} S_{22}|}{\left(1 - |S_{11}|^2 - |S_{21}|^2\right)\left(1 - |S_{22}|^2 - |S_{12}|^2\right)}, \tag{7}$$

$$DG = 10\sqrt{1 - |ECG|^2}, \tag{8}$$

$$RC = \frac{\sqrt{(S_{11} + S_{12})^2 + (S_{21} + S_{22})^2}}{\sqrt{2}}, \tag{9}$$

Measured and simulation results were compared for the reflection and transmission coefficient of the designed antenna.

4.3 Different feeding techniques used

Elsadek and Nashaat [98] investigated a new configuration of multiband/ultra-wideband antenna which has a V-shaped patch with unequal arms coupled electromagnetically to single-feed isosceles triangular PIFA thorough two unequal slots. Gayathri et al. [99] proposed a microstrip-fed planar ultra-wideband monopole antenna with band-notch characteristics and consist of an inverted cone as the radiating patch and a tapered ground plane which operates over an extremely wideband of 3–16 GHz. Mandal and Das [100] proposed regular hexagonal monopole antenna fed by a microstrip line for ultra-wide bandwidth. Peng Gao et al. [101] proposed an inverted U-shaped slot on the ground plane and a radiation patch similar to the slot that is fed by a 50-Ω microstrip line. Gautam et al. [102] proposed a novel CPW-fed compact inverted L-strip UWB microstrip antenna operating bandwidth of 2.6–13.04 GHz. Azim et al. [103] designed a microstrip line-fed planar antenna with dual-notched bands and prototyped for UWB communication applications. Telsang and Kakade [104] proposed a compact ultra-wideband microstrip patch antenna fed by a CPW and a microstrip line. Emadian and Ahmadi-Shokouh [33] proposed a very small CPW-fed rectangular slot antenna with dual-bandnotched characteristics for super UWB applications. Patre and Singh [105] demonstrated the simulation and experimental studies on a novel compact and broadband CPW-fed flower-shaped microstrip patch antenna. Singhal et al. [106, 107] presented a compact third-iteration inner-tapered tree-shaped fractal antenna for ultra-wideband applications and found that the bandwidth is enhanced by using CPW ground plane and increasing the number of iterations.

4.4 Stacked designs

Matin et al. [108] presented an antenna consisting of a U-slotted rectangular microstrip patch stacked with another patch of a different size on a separate layer and investigated its performance. Kaur et al. [109] proposed an aperture-coupled stacked Sierpinski gasket fractal antenna with a defected ground structure for UWB and wireless local area network applications.

	Dual-band	Broadband	Ultra-wideband
Bandwidth	Having two or more bands (narrow bands 5–6% bandwidth) [7–18]	Single-band having bandwidth is greater than 20% [3, 56–61]	Single-band bandwidth is greater than 109.4% [79–85]
Antenna required	Single antenna can be used as transmission and reception	Dual antennas are required for transmission and reception	Dual antennas are required for transmission and reception
Designs of antennas	Easier to achieve dual-band (means any one of the above techniques can be sufficient to achieve dual-band) [10, 11, 14, 17]	More than one design techniques required to achieve broadband [66, 68, 72, 73]	Two or more different techniques are required to achieve ultra-wideband [110]
Applications	GSM, Wi-Fi, WiMAX, DCS, 2G, 3G, Bluetooth [15, 20–22, 26]	Depending on band achieved, GSM, Wi-Fi, WiMAX, DCS, 2G, 3G, Bluetooth [59, 66]	Satellite communication, DCS, Wi-Fi, WiMax [79–85]

Table 1.
Summary of dual-band, broadband, and ultra-wideband antennas.

4.5 Parasitic designs

Ojaroudi et al. [110] proposed an inverted T-shaped strip, and by embedding a pair of T-shaped parasitic structures in the ground plane, additional resonances at the higher band are excited which provides a wide usable fractional bandwidth of more than 135%. **Table 1** summarizes the review of dual-band, broadband, and ultra-wideband antennas in this chapter.

5. Conclusions

It is observed from the review of the papers that all design of patch antennas can be converted to its equivalent circuits with basic concept high-frequency designing circuits. The proper design of circuit for particular antenna can give the clear pictures of resonating frequencies and reflection coefficient. Using circuit theory of basic structures of patch antennas, the designer can propose there circuit that will be helpful in the analysis of antenna. Further comparison of dual-band [22–55], broadband [3, 57–78], and ultra-wideband [79–110] is present in terms of application, size, bandwidth, and design. Further, it was observed that dual-band and multiband antenna is achieved through microstrip line and coaxial fed, slots, and notch structure. The ultra-wideband was achieved through coplanar structures and defected ground structures.

Acknowledgements

The authors would like to thank the Nitte Education Trust for providing the research grant no. Res/NMAMIT/03.

Conflict of interest

The authors declare no conflict of interest.

Author details

Ashish Singh[1*], Krishnananda Shet[1], Durga Prasad[1], Akhilesh Kumar Pandey[2*] and Mohammad Aneesh[3]

1 Department of Electronics and Communication Engineering, NMAM Institute of Technology, Nitte, Udupi, Karnataka, India

2 Department of Electronics Engineering, Institute of Engineering and Rural Technology, Prayagraj, Uttar Pradesh, India

3 Department of Electronics and Communication Engineering, Veer Bahadur Singh Purvanchal University, Jaunpur, Uttar Pradesh, India

*Address all correspondence to: ashsin09@rediffmail.com and akhileshpandey@iert.ac.in

References

[1] Garg R, Bhartia P, Bahl I, Ittipiboon A. Handbook of Microstrip Antennas Design. Norwood: Artech House; 2001

[2] Deschamps GA. Microstrip Microwave Antennas. In: 3rd USAF Symposium on Antennas. 1953

[3] Long SA, Walton MD. A dual-frequency stacked circular-disc antenna. IEEE Transactions on Antennas and Propagation. 1979;**27**:270-273. DOI: 10.1109/TAP.1979.1142078

[4] Derneryd KI. Broadband microstrip antenna element and array. IEEE Transactions on Antennas and Propagation. 1981;**29**:140-141. DOI: 10.1109/TAP.1981.1142530

[5] Zhao C-D. Analysis on the properties of a coupled planar dipole UWB antenna. Antennas and Wireless Propagation Letters, IEEE. 2004;**3**: 317-320. DOI: 10.1109/LAWP.2004.839242

[6] Wang B, Lo Y. Microstrip antennas for dual frequency operation. IEEE Transactions on Antennas and Propagation. 1984;**32**:938-934. DOI: 10.1109/TAP.1984.1143459

[7] Daniel E, Shevgoankar KR. Slot loaded rectangular microstrip antenna for tunable dualband operation. Microwave and Optical Technology Letters. 2005;**44**:441-444. DOI: 10.1002/mop.20661

[8] Wu WJ. 2.4/5-GHz dualband triangular slot antenna with compact operation. Microwave and Optical Technology Letters. 2005;**45**:81-84. DOI: 10.1002/mop.20730

[9] Vishvakarma S, Vishvakarma RB. Analysis of inclined slot loaded patch for dualband operation. Microwave and Optical Technology Letters. 2006;**48**:2436-2441. DOI: 10.1002/mop.21965

[10] Lim HS, Jung HC, Kim YS, Hoon N. Analysis and modeling of dualband ACMPA with asymmetric crossed slots. Microwave and Optical Technology Letters. 2011;**53**:679-686. DOI: 10.1002/mop.25761

[11] Ansari JA, Mishra A, Yadav PN, Singh P, Vishvakarma RB. Compact Triple U-slot loaded circular disk patch antenna for GSM mobile phone and optical communications. International Journal of Microwave and Optical Technology. 2011;**6**:91-99. IJMOT-2010-11-30

[12] Lu H-J, Liu H-Y. Novel dual-band design of planar slot array antenna for 4G LTE/WiMAX access points. Microwave and Optical Technology Letters. 2012;**54**:1193-1196. DOI: 10.1002/mop.26764

[13] Luo Y, Yin Y-Z, Guo Y-Y, Zhao Y. A dual-wideband antenna with an open L-slot for WLAN/WiMAX applications. Microwave and Optical Technology Letters. 2012;**54**:1499-1502. DOI: 10.1002/mop.26852

[14] Ansari JA, Mishra A, Yadav NP, Singh P, Vishvakarma BR. Analysis of W-slot loaded patch antenna for dualband operation. AEU—International Journal of Electronics and Communications. 2012;**66**:32-38. DOI: 10.1016/j.aeue.2011.04.011

[15] Gupta A, Abegaonkar MP, Basu A, Koul SK. A tunable high impedance surface and its application to dual-band microstrip antenna. Progress in Electromagnetics Research C. 2013; **43**:231-246. DOI: 10.2528/PIERC13080106

[16] Mohammad S, Nezhad A, Hassani HR, Foudazi A. A dual-band

WLAN/UWB printed wide slot antenna for MIMO/diversity applications. Microwave and Optical Technology Letters. 2013;**55**:461-465. DOI: 10.1002/mop.27391

[17] Chakraborty U, Kundu A, Chowdhury SK, Bhattacharjee AK. Compact dual-band microstrip antenna for IEEE 802.11a WLAN Application. Antennas and Wireless Propagation Letters, IEEE. 2014;**13**:407-410. DOI: 10.1109/LAWP.2014.2307005

[18] Tsai L-C. A triple-band bow-tie-shaped CPW-fed slot antenna for WLAN applications. Progress in Electromagnetics Research C. 2014;**47**:167-171. DOI: 10.2528/PIERC14011002

[19] Liu WC, Chen WR. CPW-fed compact meandered patch antenna for dualband operation. Electronics Letters. 2004;**40**:1094-1095. DOI: 10.1049/el:20045335

[20] Sung Y, Ahn CS, Kim Y-S. Microstripline fed dual frequency dielectric resonator antenna. Microwave and Optical Technology Letters. 2004;**42**:388-390. DOI: 10.1002/mop.20313

[21] Su W, Row JS, Wu JF. Design of a single feed dual frequency microstrip antenna. Microwave and Optical Technology Letters. 2005;**47**:114-116. DOI: 10.1002/mop.21096

[22] Chen SY, Hsu P. Broadband radial slot antenna fed by coplanar waveguide for dual frequency operation. IEEE Transactions on Antennas and Propagation. 2005;**53**:3448-3452. DOI: 10.1109/TAP.2005.858574

[23] Sze JY, Hsu CIG, Jiao JJ. CPW fed circular slot antenna with slit back-patch for 2.4/5 GHz dualband operation. Electronics Letters. 2006;**42**:563-564. DOI: 10.1049/el:20064479

[24] Huang Y, Lin HC, Kuo JS. Dualband monopole antenna excited by a capacitive coupling feed for WLAN applications. Microwave and Optical Technology Letters. 2007;**49**:1135-1138. DOI: 10.1002/mop.22380

[25] Yoon W, Lee J, Kang SP, Kang SY, Lee HC, Park HD. A planar CPW-fed slot antenna on thin substrate for dualband operation of WLAN applications. Microwave and Optical Technology Letters. 2009;**51**:2799-2802. DOI: 10.1002/mop.24742

[26] Lin WC, Wu CM, Chu NC. A compact CPW fed slotted patch antenna for dualband. Antennas and Wireless Propagation Letters, IEEE. 2010;**9**:110-113. DOI: 10.1109/LAWP.2010.2044135

[27] Singh A, Aneesh M, Kamakshi K, Ansari JA. Circuit theory analysis of aperture coupled patch antenna for wireless communication. Radioelectronics and Communications Systems. 2018;**61**(4):168-179. DOI: 10.3103/S0735272718040040

[28] Singh A, Aneesh M, Kamakshi K, Ansari JA. Analysis of microstrip line fed patch antenna for wireless communications. Open Engineering Journal. 2017;**7**:279-286. DOI: 10.1515/eng-2017-0034

[29] Gulam Nabi Alsath M, Sridhar B, Malathi K, Rajesh Kumar R, Karthik N, Henridass A. A dual band frequency and pattern reconfigurable dielectric resonator antenna. Progress In Electromagnetics Research C. 2012;**27**:115-128. DOI: 10.2528/PIERC11121406

[30] Li Y, Li W, Ye Q. A compact asymmetric coplanar strip-fed dual-band antenna for 2.4/5.8 GHz WLAN applications. Microwave and Optical Technology Letters. 2013;**55**:2066-2070. DOI: 10.1002/mop.27741

[31] Moosazadeh M, Esmati Z. Small planar dual-band microstrip-fed monopole antenna for wireless local

area network applications using slotted conductor-backed plane. Microwave and Optical Technology Letters. 2013; **55**:2380-2383. DOI: 10.1002/mop.27814

[32] Chen S-W, Wang D-Y, Tu W-H. Dual-band/tri-band/broadband CPW-fed stepped-impedance slot dipole antennas. IEEE Transactions on Antennas and Propagation. 2014;**62**: 485-490. DOI: 10.1109/TAP.2013.2287523

[33] Emadian SR, Ahmadi-Shokouh J. Very small dual band-notched rectangular slot antenna with enhanced impedance bandwidth. IEEE Transactions on Antennas and Propagation. 2015;**63**:4529-4534. DOI: 10.1109/TAP.2015.2456905

[34] Singh A, Ansari JA, Kamakshi K, Aneesh M, Sayeed SS. Analysis of slot loaded compact patch antennas for dualband operation. International Journal of Applied Electromagnetics and Mechanics. 2015;**47**:163-175. DOI: 10.3233/JAE-140020

[35] Karmakar NC. Shorting strap tunable stacked patch PIFA. IEEE Transactions on Antennas and Propagation. 2004;**52**:2877-2884. DOI: 10.1109/TAP.2004.835124

[36] Pioch S, Laheurte JM. Low profile dualband antenna based on a stacked configuration of EBG and plain patches. Microwave and Optical Technology Letters. 2005;**44**:207-209. DOI: 10.1002/mop.20589

[37] Kim YD, Kim HY, Lee HM. Dual-band LTCC chip antenna design using stacked meander patch for mobile handsets. Microwave and Optical Technology Letters. 2005;**45**:271-273. DOI: 10.1002/mop.20793

[38] Anguera J, Puente C, Borja C, Soler J. Dual-frequency broadband-stacked microstrip antenna using a reactive loading and fractal-shaped radiating edge. IEEE Antennas and Wireless Propagation Letters. 2007;**6**: 309-312. DOI: 10.1109/LAWP.2007.891523

[39] Ansari JA, Singh P, Dubey SK, Khan RU, Vishvakarma BR. H-shaped stacked patch antenna for dualband operation. Progress in Electromagnetics Research B. 2008;**5**:291-302. DOI: 10.2528/PIERB08031203

[40] Ma S-L, Row J-S. Design of single-feed dual-frequency patch antenna for GPS and WLAN applications. IEEE Transactions on Antennas and Propagation. 2011;**59**:3433-3436. DOI: 10.1109/TAP.2011.2161453

[41] Benkouda S, Amir M, Fortaki T, Benghalia A. Dual-frequency behavior of stacked high T_c superconducting microstrip patches. Journal of Infrared, Millimeter, and Terahertz Waves. 2011; **32**:1350-1366. DOI: 10.1007/s10762-011-9842-1

[42] Batgerel A, Eom S-Y. Dual-band microstrip antenna structure combined with high-gain dielectric rod and sleeve-dipole elements. Microwave and Optical Technology Letters. 2012;**54**:1835-1838. DOI: 10.1002/mop.26933

[43] Mishra B, Singh V, Singh R. Dual and Wide band slot loaded stacked microstrip patch antenna for WLAN/WIMAX application. Microsystem Technologies. 2017;**23**:3467-3475. DOI: 10.1007/s00542-016-3120-z

[44] Anguera J, Fractus SA, Martinez E, Puente C, Borja C. Broadband dual frequency microstrip patch antenna with modified sierpinski fractal geometry. IEEE Transactions on Antennas and Propagation. 2004;**52**: 66-73. DOI: 10.1109/TAP.2003.822433

[45] Chen WH, Feng ZH, Fan MY, Furuya Y. Dualband reconfigurable antenna for wireless communication. Microwave and Optical Technology

Letters. 2004;**40**:503-505. DOI: 10.1002/mop.20016

[46] Cho YJ, Hwang SH, Park SO. A dual-band internal antenna with a parasitic patch for mobile handsets and the consideration of the handset case and battery. Antennas and Wireless Propagation Letters, IEEE. 2005;**4**: 429-432. DOI: 10.1109/LAWP.2005. 859388

[47] Liu L, Zhu S, Langley R. Dualband triangular patch antenna with modified ground plane. Electronics Letters. 2007; **43**:140-141. DOI: 10.1049/el:20073643

[48] Peng L, Ruan C-L. A microstrip fed patch antenna with two parasitic invert l stubs for dual-band WLAN applications. Wireless Personal Communications. 2011;**57**(4):727-734. DOI: 10.1007/ s11277-009-9873-2

[49] Shivnarayan, Vishvakarma BR. Analysis of dualband patch antenna for mobile communications. Microwave and Optical Technology Letters. 2005; **47**:558-564. DOI: 10.1002/mop.21229

[50] Deshmukh A, Kumar G. Even-mode multiport network model for slotted dualband rectangular microstrip antennas. Microwave and Optical Technology Letters. 2006;**48**:798-804. DOI: 10.1002/mop.21479

[51] Mishra A, Singh P, Yadav NP, Ansari JA, Vishvakarma BR. Compact shorted microstrip patch antenna for dual- band operation. Progress in Electromagnetics Research C. 2009;**9**: 171-182. DOI: 10.2528/PIERC09071007

[52] Thomas KG, Sreenivasan M. Compact CPW fed dualband antenna. Electronics Letters. 2010;**46**:13-14. DOI: 10.1049/el.2010.1729

[53] Lin D-B, Tang I-T, Wei Y-J. Compact dual-band-notched CPW-fed wide-slot antenna for WLAN and WiMAX applications. Microwave and Optical Technology Letters. 2011;**53**: 1496-1501. DOI: 10.1002/mop.26042

[54] Singh A, Ansari JA, Kamakshi K, Mishra A, Aneesh M. Compact notch loaded half disk patch antenna for dualband operation. Annals of Telecommunication. 2014;**69**:475-483. DOI: 10.1007/s12243-013-0383-6

[55] Mishra A, Ansari JA, Singh P, Yadav NP, Vishvakarma BR. L-strip feed circular disk dual resonator patches antenna for wireless communication. Wireless Personal Communications. 2013;**72**:795-807. DOI: 10.1007/ s11277-013-1316-4

[56] Verma S, Ansari JA, Singh A. Truncated equilateral triangle microstrip patch with and without substrate. An International Journal of Wireless Personal Communication. 2017;**95**:873-889. DOI: 10.1007/ s11277-016-3803-x

[57] Hall PS, Wood C, Garrett C. Wide-bandwidth microstrip antennas for circuit integration. Electronics Letters. 1979;**15**:558-560. DOI: 10.1049/el: 19790329

[58] Mestdagh S, Raedt WD, Vandenbosch GAE. CPW-fed stacked microstrip antennas. IEEE Transactions on Antennas and Propagation. 2004; **52**(1):74-83. DOI: 10.1109/TAP.2003. 822407

[59] Yang SSL, Luk KM. Wideband folded-patch antennas fed by L-shaped probe. Microwave and Optical Technology Letters. 2005;**45**(4):352-355. DOI: 10.1002/mop.20821

[60] Ge Y, Esselle KP, Bird TS. A wideband probe-fed stacked dielectric resonator antenna. Microwave and Optical Technology Letters. 2006;**48**(8): 1630-1633. DOI: 10.1002/mop.21716

[61] Zhang L-N, Zhong S-S, Xu S-Q. Broadband U-shaped dielectric

resonator antenna with elliptical patch feed. Electronics Letters. 2008;**44**(16): 947-949. DOI: 10.1049/el:20081253

[62] Natrajan V, Chettier E, Chatterjee D. An ultra-wideband dual-stacked U-slot microstrip antenna. IEEE Antennas and Propagation Society International Symposium. 2004;**3**: 2939-2942. DOI: 10.1109/APS.2004.1331994

[63] Lui HW, Luk KM. Wideband stacked patch antenna fed by meandering probe. Electronics Letters. 2005;**41**(6):297-298. DOI: 10.1049/el:20057671

[64] Ghannaum H, Baries S, Roblin C. Probe fed stacked patch antenna for ultra-wideband. In: IEEE International Conference Ultra-wideband, ICU 2005. 2005. pp. 97-102. DOI: 10.1109/ICU.2005.1569965

[65] Deshmukh AK, Kumar G. Compact broadband stacked microstrip antennas. In: IEEE Antennas and Propagation Society International Symposium. 2006. pp. 3747-3750. DOI: 10.1109/APS.2006.1711437

[66] Wu J-W, Hsiao H-M, Lu J-H, Chang S-H. Dual broadband design of rectangular slot antenna for 2.4 and 5 GHz wireless communication. Electronics Letters. 2004; **40**(23):1461-1463. DOI: 10.1049/el:20046873

[67] Deshmukh AA, Kumar G. Broadband pairs of slot-loaded rectangular microstrip antennas. Microwave and Optical Technology Letters. 2005;**47**(3):223-226. DOI: 10.1002/mop.21130

[68] Deshmukh AA, Kumar G. Various slots loaded broadband and compact circular microstrip antennas. Microwave and Optical Technology Letters. 2006;**48**(3):435-439. DOI: 10.1002/mop.21373

[69] Yao F-w, Zhong S-S. Broadband and high gain microstrip slot antenna. Microwave and Optical Technology Letters. 2006;**48**(11):2210-2212. DOI: 10.1002/mop.21907

[70] Chen W-F, Tsai C-C, Huang C-Y. Compact wide-slot antenna for ultrawideband communications. Electronics Letters. 2008;**44**(15): 892-893. DOI: 10.1049/el:20081563

[71] Mo L, Zhang H, Zhou H. Broadband UHF RFID tag antenna with a pair of U slots mountable on metallic objects. Electronics Letters. 2008;**44**(20): 1173-1174. DOI: 10.1049/el:20089813

[72] Aneesh M, Singh A, Kamakshi K, Ansari JA. Performance investigations of S-shaped RMSA using multilayer perceptron neural network for S-band applications. Radioelectronics and Communications Systems. 2019;**62**: 400-408. DOI: 10.3103/S0735272 71908003X

[73] Aneesh M, Siddiqui GM, Ansari JA, Singh A, Kamakshi K. Inset feed toppled H-shapped microstrip patch antenna for PCS/WiMax application. TELKOMNIKA Indonesian Journal of Electrical Engineering and Computer Science. 2016;**1**(2):365-370. DOI: 10.11591/ijeecs.v1.i2.pp365-370

[74] Liu Z, Boyle K. Bandwidth enhancement of a quarter-wavelength slot antenna by capacitive loading. Microwave and Optical Technology Letters. 2009;**51**(9):2114-2116. DOI: 10.1002/mop.24579

[75] Tang ZJ, He Y-G. Broadband microstrip antenna with U and T slots for 2.45/2.41 GHz. RFID tag. Electronics Letters. 2009;**45**(18):926-928. DOI: 10.1049/el.2009.0387

[76] Kimouche H, Abed D, Atrouz B, Aksas R. Bandwidth enhancement of rectangular monopole antenna using modified Semi-elliptical ground plane

and slots. Microwave and Optical Technology Letters. 2009;**52**(1):54-58. DOI: 10.1002/mop.24830

[77] Liu H-W, Chia-Hao K, Wang T-S, Yang C-F. Compact monopole antenna with band-notched characteristic for UWB applications. Antennas and Wireless Propagation Letters, IEEE. 2010;**9**:397-400. DOI: 10.1109/LAWP.2010.2049633

[78] Ghaderi MR, Mohajeri F. A compact hexagonal wide-slot antenna withmicrostrip-fed monopole for UWB application. Antennas and Wireless Propagation Letters, IEEE. 2011;**10**:682-685. DOI: 10.1109/LAWP.2011.2158629

[79] Ojaroudi N, Ojaroudi M. Novel design of dual band-notched monopole antenna with bandwidth enhancement for UWB applications. Antennas and Wireless Propagation Letters, IEEE. 2013;**12**:698-701. DOI: 10.1109/LAWP.2013.2264713

[80] Gao P et al. Compact printed wide-slot UWB antenna with 3.5/5.5-GHz dual bandnotched characteristics. Antennas and Wireless Propagation Letters, IEEE. 2013;**12**:983-986. DOI: 10.1109/LAWP.2013.2277591

[81] Shinde PN, Mishra BK. Compact thin ground plane UWB antenna with dual band stop characteristics. Microwave and Optical Technology Letters. 2013;**55**:1045-1049. DOI: 10.1002/mop.27470

[82] Gong B et al. Compact slot antenna for ultra-wide band applications. Microwaves, Antennas & Propagation, IET. 2014;**8**:200-205. DOI: 10.1049/iet-map.2013.0067

[83] Haraz OM, Sebak A-R, Alshebeili SA. Design of ultrawideband multilayer slot coupled vertical microstrip transitions employing novel patch shapes. Microwave and Optical Technology Letters. 2015;**57**(3):747-756. DOI: 10.1002/mop.28940

[84] Shameena VA et al. Compact ultra-wideband planar serrated antenna with notch band ON/OFF control. Electronics Letters. 2006;**42**:1323-1324. DOI: 10.1049/el:20062863

[85] Ojaroudi M, Urmia I, Ghanbari G, Ojaroudi N, Ghobadi C. Small square monopole antenna for UWB applications with variable frequency band-Notch function. Antennas and Wireless Propagation Letters, IEEE. 2009;**8**:1061-1064. DOI: 10.1109/LAWP.2009.2030697

[86] Movahedinia R, Azarmanesh MN. Ultra-wideband band-notched printed monopole antenna. Microwaves, Antennas & Propagation, IET. 2010;**4**:2179-2186. DOI: 10.1049/iet-map.2009.0638

[87] Ojaroudi M, Yazdanifard S, Ojaroudi N, Sadeghzadeh RA. Band-Notched small square-ring antenna with a pair of T-shaped strips protruded inside the square ring for UWB applications. Antennas and Wireless Propagation Letters, IEEE. 2011;**10**:227-230. DOI: 10.1109/LAWP.2011.2131630

[88] Li L, Zhou Z-L, Hong J-S, Wang B-Z. Compact dual-band-notched UWB planar monopole antenna with modified SRR. Electronics Letters. 2011;**47**:950-951. DOI: 10.1049/el.2011.1874

[89] Ojaroudi N, Ojaroudi M, Ghadimi N. UWB omnidirectional square monopole antenna for use in circular cylindrical microwave imaging systems. Antennas and Wireless Propagation Letters, IEEE. 2012;**11**:1350-1353. DOI: 10.1109/LAWP.2012.2227137

[90] Mehranpour M, Nourinia J, Ghobadi C, Ojaroudi M. Dual band-notched square monopole antenna for

ultrawideband applications. Antennas and Wireless Propagation Letters, IEEE. 2012;**11**:172-175. DOI: 10.1109/LAWP.2012.2186552

[91] Zhu F et al. Multiple band-notched UWB antenna with band-rejected elements integrated in the feed line. IEEE Transactions on Antennas and Propagation. 2013;**61**:3952-3960. DOI: 10.1109/TAP.2013.2260119

[92] Mandal T, Das S. UWB printed plaque monopole antennas for tri-band rejection. Microwave and Optical Technology Letters. 2013;**55**:674-680. DOI: 10.1002/mop.27359

[93] Natarajamani S, Behera SK, Patra SK. A compact planar diversity antenna for ultra-wideband application with band-notched function. Microwave and Optical Technology Letters. 2013;**55**:758-762. DOI: 10.1002/mop.27435

[94] Karimian R, Oraizi H, Fakhte S. Design of a compact ultra-wide-band monopole antenna with band rejection characteristics. Microwaves, Antennas & Propagation, IET. 2014;**8**:604-610. DOI: 10.1049/iet-map.2013.0085

[95] Shou Hui Zheng X, Liu Y, Tentzeris MM. Optically controlled reconfigurable band-notched UWB antenna for cognitive radio systems. Electronics Letters. 2014;**50**:1502-1504. DOI: 10.1049/el.2014.2226

[96] Rajesh Kumar V, Raghavan S. Bandwidth enhanced compact fractal antenna for UWB applications with 5–6 GHz band rejection. Microwave and Optical Technology Letters. 2015;**57**:607-613. DOI: 10.1002/mop.28913

[97] Gautam AK, Yadav S, Rambabu K. Design of ultra compact UWB antenna with band notch characteristics for MIMO applications. IET Microwaves Antennas & Propagation. 2018;**12**:1895-1900. DOI: 10.1049/iet-map.2018.0012

[98] Elsadek H, Nashaat DM. Multiband and UWB V-shaped antenna configuration for wireless communications applications. Antennas and Wireless Propagation Letters, IEEE. 2008;**7**:89-91. DOI: 10.1109/LAWP.2007.900953

[99] Gayathri R et al. Band-notched inverted-cone monopole antenna for compact UWB systems. Electronics Letters. 2008;**44**:1170-1171. DOI: 10.1049/el:20081770

[100] Mandal T, Das S. Ultrawideband-printed hexagonal monopole antennas with WLAN band rejection. Microwave and Optical Technology Letters. 2012;**54**:1520-1525. DOI: 10.1109/IndianAW.2011.6264945

[101] Gao P et al. Compact printed wide-slot UWB antenna with 3.5/5.5-GHz dual band notched characteristics. Antennas and Wireless Propagation Letters, IEEE. 2013;**12**:983-986. DOI: 10.1109/LAWP.2013.2277591

[102] Gautam K, Yadav S, Kanaujia BK. A CPW-FED compact inverted L-strip UWB microstrip antenna. Microwave and Optical Technology Letters. 2013;**55**:1584-1589. DOI: 10.1002/mop.27651

[103] Azim R, Islam MT, Mobashsher AT. Dual band-notch UWB antenna with single tri-arm resonator. Antennas and Wireless Propagation Letters, IEEE. 2014;**13**:670-673. DOI: 10.1109/LAWP.2014.2314486

[104] Telsang TM, Kakade AB. Ultra wideband slotted semicircular patch antenna. Microwave and Optical Technology Letters. 2014;**56**:362-369. DOI: 10.1002/mop.28102

[105] Patre SR, Singh SP. CPW-fed flower-shaped patch antenna for broadband applications. Microwave and

Optical Technology Letters. 2015;**57**: 2908-2913. DOI: 10.1002/mop.29480

[106] Singhal S, Goel T, Singh AK. Inner tapered tree-shaped fractal antenna for UWB applications. Microwave and Optical Technology Letters. 2015;**57**: 559-567. DOI: 10.1002/mop.28900

[107] Singhal S, Singh AK. Inner Tapered Tree-Shaped Ultra-Wideband Fractal Antenna with Polarization Diversity. Rijeka: IntechOpen; 2015. DOI: 10.5772/intechopen.86071

[108] Matin MA, Sharif BS, Tsimenidis CC. Dual layer stacked rectangular microstrip patch antenna for ultra wideband applications. Microwaves, Antennas & Propagation, IET. 2007;**1**:1192-1196. DOI: 10.1049/iet-map:2007005

[109] Kaur A, Khanna R, Kartikeyan MV. A stacked sierpinski gasket fractal antenna with a defected ground structure for UWB/WLAN/RADIO astronomy/STM Link applications. Microwave and Optical Technology Letters. 2015;**57**:2786-2792. DOI: 10.1002/mop.29442

[110] Ojaroudi N, Ojaroudi Y, Ojaroudi S. Compact ultra-wideband monopole antenna with enhanced bandwidth and dual band-stop properties. International Journal of RFand Microwave Computer-Aided Engineering. 2015;**25**:346-357. DOI: 10.1002/mmce.20868